The Common Information Model CIM

Mathias Uslar, Michael Specht,
Sebastian Rohjans, Jörn Trefke,
and José Manuel Vasquez González

The Common Information Model CIM

IEC 61968/61970 and 62325 – A Practical Introduction to the CIM

Springer

Authors

Dr.-Ing. Mathias Uslar
OFFIS - Institut für Informatik
Escherweg 2
26131 Oldenburg
E-mail: uslar@offis.de

Dipl.-Inf. (FH) Michael Specht
OFFIS - Institut für Informatik
Escherweg 2
26131 Oldenburg
E-mail: michael.specht@offis.de

Dipl.-Inform. Sebastian Rohjans
OFFIS - Institut für Informatik
Escherweg 2
26131 Oldenburg
E-mail: sebastian.rohjans@offis.de

Dipl.Inf. Jörn Trefke
OFFIS - Institut für Informatik
Escherweg 2
26131 Oldenburg
E-mail: Joern.trefke@offis.de

Dipl.-Wirt.Inf. José Manuel Vasquez González
OFFIS - Institut für Informatik
Escherweg 2
26131 Oldenburg
E-mail: jose.gonzalez@offis.de

ISBN 978-3-642-25214-3
e-ISBN 978-3-642-25215-0

DOI 10.1007/978-3-642-25215-0

Library of Congress Control Number: 2011942873

Typeset by Scientific Publishing Services Pvt. Ltd., Chennai, India.

Printed on acid-free paper

9 8 7 6 5 4 3 2 1

springer.com

"The only reason for time is so that everything doesn't happen all at once."
Albert Einstein

Foreword

Within the smart grid, the combination of automation equipment, communication technology and IT is crucial. Interoperability of devices and systems can be seen as the key enabler of smart grids. Therefore, international initiatives have been started in order to identify interoperability standards for smart grids. A prominent example is the European mandate M/490, which develops a reference architecture, a first set of smart grid applicable standards and use cases. As the European initiative has an ambitious schedule, relying on existing accepted standards and solutions seems most appropriate.

IEC 62357, the so called Seamless Integration Architecture, is one of these very core standards, which has been identified by recent smart grid initiatives and roadmaps to be essential for building and managing intelligent power systems.. It is fully applicable in the power utilities domain, globally accepted, future proof and can be extended to multi-utility. From the operative and strategic perspective of network operations, it is the reasonable choice for a basis architecture. The Seamless Integration Architecture provides an overview of the interoperability and relations between further standards from IEC TC 57 - the two prominent standards families IEC 61850: Communication networks and systems in power utility automation and the IEC 61970/61968: Common Information Model.

Both of them have proven to be mature standards for interface interoperability and engineering; consequently, they are the cornerstone of the IEC Smart Grid Standardization Roadmap. Within the IEC TC 57 "Power system management and associated information exchange", both standards are in the focus of experts' work. IEC 61850 and its sub-parts have already gained a lot of momentum. IEC 61850 has become a global success story for the power automation industry, having currently worldwide several hundred thousand devices and systems in operation. As IEC 61850 is an established global standard, a lot of literature is available, and there exist tutorials and trainings.

And now comes the opportunity for CIM. Although a lot of implementations are in place, few textbooks and introduction material are available. This book tries to bridge the existing gap - providing utilities, vendors, regulators and all interested readers a chance to get to know the CIM. The authors from OFFIS have designed this book to be an introduction to ICT and CIM for power engineers, giving an introduction to SCADA/EMS and DMS for ICT engineers and an overview for decision makers in utilities. As IEC TC 57 secretary, I fully welcome this first textbook on the CIM, explaining this important work of IEC TC57.

I hope that a lot of readers will benefit from this textbook.

Nuremberg, August 2011 *Dr. Heiko Englert*
IEC TC 57 and CLC 57 secretary, SIEMENS AG

Foreword

The Common Information Model (CIM) began with an EPRI (Electric Power Research Institute) project in the nineties. The project was named CCAPI (Control Center Application Programming Interfaces). At that time, Unified Modeling Language, Object Oriented Languages, internet associated technologies were not as widespread as they are nowadays.

The key idea of CIM is to define a common language in order to facilitate message exchanges. The analogy is the same as the choice of English for supporting such a book. English is chosen because nowadays most people use it in their international business exchanges.

In the mid-nineties, CIM was handed over to IEC (International Electrotechnical Commission), and specifically a new group in IEC TC57 was created: WG 14. Its scope was to define System Interfaces for Distribution Management. EDF Research and Development Division began its work in these IEC TC57 CIM oriented working groups at this time.

In the Network Operation Domain, some utilities, like EDF, were renovating their Remote Control Systems (named also SCADA for Supervisory Control And Data Acquisition). At that time EDF R&D began to use Object Oriented Techniques and was developing SCADA-related software using C++ and CORBA (Common Object Request Broker Architecture) to exchange information. Java was in its infancy and XML was not yet known in the utility domain as they are now.

EDF R&D really was able to promote CIM usage in our utility when a three-year project, called Cimergy, was funded in 2004. Based on former ETSO association work, we were able to associate a methodology the UN-CEFACT Core Component Specification to the CIM model, which facilitated CIM usage. At the same time we also developed several CIM API (Application Programming Interfaces) for different Tools which were not inter-operable. Doing that, we were able to participate in interoperability tests funded by EPRI since 2000, and now it is also funded by the UCA association under which CIM User group was created in 2005. These interoperability tests and UCA CIM user group helped us to better understand what CIM was and how it was used by vendor products and to promote CIM usage at the Distribution level.

In Europe, 2009 was a cornerstone year with the adoption of CIM by ENTSO-E (European Networks Transmission Operators - Electricity) as its new data exchange format for insuring the reliability of the European Transmission Networks. Several network operation and planning-oriented products are becoming CIM-compliant. Since then, EDF operational divisions and many other utilities have started different projects using CIM. During this time, worldwide experts promoted CIM usage in different project: European R&D projects, Smartgrid demonstrators involving utilities, in-house utility projects, etc.

With Smartgrids, Electrical Networks need to leverage Information and Communication technologies (ICT). With Smartgrid, interoperability is a key word. CIM and other standards like 61850 for system automation, DLMS-COSEM for smart metering will help to increase interoperability. We firmly believe that a Model Engineering Approach is a valuable solution in reaching this interoperability objective.

Nevertheless, CIM usage is not sufficient in the Utility area, that is why such a book is a piece of the puzzle. It will facilitate the harmonization of CIM with other relevant standards. That is why I'd like to congratulate the authors for writing this book: educating people, from students to regulators and other stakeholders, about the CIM international standard and related technologies. This education is a "must". I'm sure this book will help many people as it will help them to de-mystify CIM and its usages.

I'd like also to thank all worldwide experts who contributed and still contribute to improving CIM and who indirectly contributed to this reference book.

All the best to this first edition!

Paris, Septembre 2011

Eric Lambert
EDF R&D Project Manager
IEC/CLC TC57 Member
UCA Executive Committee

Preface

The scope of this book has been developed by the authors with the focus on addressing more than one particular audience. The overall topic of the book is the Common Information Model CIM as originally designed by the IEC and its theoretical foundations which are covered in Part 1 of this book from chapters 1 to 4. Additionally, part 2 with the chapters 5 to 8 of the book strongly focuses on the direct utilization, application and tooling for the CIM.

The overall objectives of the authors for this first edition of the book are to provide a useful companion for power system engineers getting an introduction to CIM, ICT developers to get to know the background of the CIM modeling scope where standard ICT technologies like UML are introduced. Furthermore, decision makers in utilities or vendors trying to find out on improving architectures or integration costs through facilitating the use of CIM are adressed. As those viewpoints differ from each other, we try to provide useful information to all mentioned audiences by setting no extreme focus on one particular aspect, e.g. basics or application. Through dividing the book into two large parts, we want to achieve the following objectives and provide a well-weighted overview:

- to transfer the know-how about the technical, functional and economic aspects around the CIM to decision makers for SCADA/EMS/DMS and secondary IT as well as application landscapes for utilities
- to make power system engineers aware of the new ICT based technologies for service-oriented architectures, ESB and domain modeling techniques like UML, XML and RDF which have to be used in context with the CIM's serializations
- to provide existing CIM users with a better knowledge on related use cases, tools to be applied and background on the distribution of the CIM and its place in terms of the Smart Grid and IEC's Seamless Integration Architecture SIA.
- to provide a reference and meaningful starting point for further information on certain aspects which are not covered by this book (yet) and future aspects of the CIM as well as getting the user involved in standardization and CIM user's group.

The following table provides more detailed guidance for the readers with the various aforementioned backgrounds and education in order to select specific chapters which are of highest interest to complement their existing knowledge with the information on power systems modeling and data integration.

A ○ depicts the less important chapters for an audience whereas a • emphasizes a special benefit for this audience for reading the mentioned chapter.

Readers:	1	2	3	4	5	6	7	8	Annex
Students	•	•	•	○	○	○	○	○	○
Beginners in CS, Power Systems, EMS	•	•	•	○	•	○	○	○	○
Decision makers in CS, Power Systems, EMS	•	○	•	○	○	○	•	•	○
Developers	○	○	•	•	•	•	○	○	○
Engineering Specialists	○	○	•	○	○	•	○	○	○
SCADA Specialists	○	•	•	•	○	○	○	○	○
Testing Specialists	○	○	•	○	○	○	•	○	○
SOA Architects	•	○	•	•	○	○	○	○	○

Oldenburg, September 2011

Mathias Uslar
Michael Specht
Sebastian Rohjans
Jörn Trefke
José Manuel Vasquez González

Acknowledgements

The authors would like to thank all the experts which contributed to this book. Without the input from the IEC working groups, especially IEC TC 57 WG 13, 14 and 16 and corrections by several experts, including our colleagues from projects a national and international level, our colleagues and friends in international and national standardization and our partners at home accepting the additional amount of work and time spent, this book would not have been written as it is.

Namely, we would like to thanks Heiko Englert from Siemens being the TC 57 secretary to provide our foreword, we would like to thank our colleagues here at OFFIS from the Energy Management group for providing input to this book, our colleagues from projects at BTC and EWE, Germany, Lars Nordström form KTH for always providing input and help during the first years here at our OFFIS CIM projects, John Gillerman from SISCO and Jean-Francois Cabadi from Alstom for a good cooperation on the OPC UA mappings, Wolfgang Mahnke from ABB for motivating us to do a proposal for Springer, Eric Lambert from EdF for constant encouragement on European CIM initiatives and various others like the TC 57 WG 14 and 13 Convenors Greg Robinson and Terry Saxton.

Contents

Acronyms

A&E	Alarms and Events
AMI	Advanced Metering Infrastructure
AMR	Automated Meter Reading
API	application programming interface
BDEW	Bundesverband der Energie- und Wasserwirtschaft e.V. (engl.: German Association of Energy and Water Industries)
BMU	Bundesministerium für Umwelt, Naturschutz und Reaktorsicherheit (engl.: German Federal Ministry of Environment, Nature Conservation and Nuclear Safety)
BMWi	Bundesministeriums für Wirtschaft und Technologie (engl.: German Federal Ministry of Economics and Technology)
BPMN	Business Process Model and Notation
CASE	Computer Aided Software Engineering
CCAPI	Control Center Application Programming Interface
CD	Committee Draft
CDPSM	Common Distribution. Power System Model
CDV	Committee Draft for Vote
CEN	European Committee for Standardization
CENELEC	European Committee for Electrotechnical Standardization
CHP	Combined Heat and Power
CIGRE	Conseil International des Grand Reseaux Electriques
CIM	Common Information Model
CIMug	CIM users group
CIS	Common Interface Specifications
COM	Component Object Model
CORBA	Common Object Request Broker Architecture
COSEM	COmpanion Specification for Energy Metering
CPSM	Common Power System Model
DA	Data Access
DAF	Data Access Facility
DAIS DA	Data Acquisition from Industrial Systems Section Data Access

DAIS A&E	Data Acquisition from Industrial Systems Alarms and Events
DER	Decentralized Energy Resources
DEWG	Data Exchange Working Group
DHTML	Dynamic HTML
DIN	Deutsches Institut für Normung e.V.
DKE	DKE Deutsche Kommission Elektrotechnik Elektronik Informationstechnik im DIN und VDE (engl.: German Commission for Electrical, Electronic & Information Technologies)
DLMS	Device Language Message specification
DMS	Distribution Management System
DoC	Department of Commerce
DoE	Department of Energy
DOM	Document Object Model
DPP	Data protection and privacy
DSO	Distribution System Operator
DTD	Document Type Definition
EAI	Enterprise Application Integration
ebXML	Electronic Business using XML
EDI	Electronic data interchange
EIIA	Enterprise Information Integration Adapter
EISA	Energy Independence and Security Act
EMS	Energy Management System
ENTSO-E	European Network of Transmission System Operators for Electricity
EPRI	Electric Power Research Institute
EPS	Electric Power System
ERCOT	Electric Reliability Council of Texas
ERP	Enterprise Resource Planning
ESB	Enterprise Service Bus
ESO	European Standardization Organizations
ETP	European Technology Platform
ETS	European Telecommunications Standards Institute
EV	Electrical Vehicle
FACTS	Flexible Alternating Current Transmission System
FDIS	Final Draft International Standard
GDA	Generic Data Access
GES	Generic Eventing and Subscription
GIS	Geographic Information System
GID	Generic Interface Definition
HAN	Home Area Network
HDA	Historical Data Access
HTML	HyperText Markup Language
HSDA	High-Speed Data Access
HVDC	High-Voltage Direct Current
ICT	Information and Communication Technologies

IDL	Interface Definition Language
IEC	International Electrotechnical Commission
IED	Intelligent Electronic Device
IEM	Information Exchange Model
IETF	Internet Engineering Task Force
IEV	International Electrotechnical Vocabulary, Electropedia
IOP	Interoperability
IRI	Internationalized Resource Identifier
IRM	Interface Reference Model
IS	International Standard
ISO	International Organization for Standardization
ITU	International Telecommunication Union
JET	Java Emitter Templates
JMS	Java Messaging System
JTC1	Joint Technical Committee 1
JWG-SG	Joint Working Group Smart Grid
LAN	Local Area Network
NC	National Committee
NDR	Naming and Design Rules
NERC	North American Electric Reliability Council
NIST	National Institute of Standards and Technology
NWIP	New Work Item Proposal
MDM	Meter Data Management
METI	Ministry of Economy, Trade and Industry
MES	Manufacturing Execution Systems
MKE	Ministry of Knowledge Economy
MT	Maintenance Team
OCL	Object Constraint Language
OMG	Object Management Group
OMS	Object Management System
OPC-UA	OPC - Unified Architecture
OWL	Web Ontology Language
PAP	Priority Action Plan
PAS	Publicly Available Specifications
PLC	Power Line Communications
PT	Project Team
PWI	Potential New Work Item Proposal
RDBMS	Relational DataBase Management System
RDF	Resource Description Framework
RDFS	RDF Schema
RNR	Resource Name Repository
RTO	Regional Transmission Organization
SDO	Standard Developing Organization
SAS	Substation Automation System
SC	Subcommittee

SCADA	Supervisory Control and Data Acquisition
SEP	Smart Energy Profile
SERA	Smart Energy Reference Architecture
SG	Strategic Group
SGCC	State Grid Corporation of China
SGIP	Smart Grid Interoperability Panel
SGML	Standard Generalized Markup Language
SGIS	Smart Grid Information Security
SIA	Seamless Integration Reference Architecture
SIDM	System Interfaces for Distribution Management
SM-CG	Smart Meters Co-Ordination Group
SMB	Standardization Management Board
SME	Small and Medium Enterprises
SOA	Service-oriented architecture
SPARQL	SPARQL Protocol and RDF Query Language
SQL	Structured Query Language
SVG	Scalable Vector Graphic
TC	Technical Committee
TCP/IP	Transmission Control Protocol and Internet Protocol
TLS	Transport Layer Security
TLY3	Typical Meteorological Year
TR	Technical Report
TS	Technical Specification
TSDA	Time Series Data Access
TSO	Transmission System Operator
TTA	Technology Trend Assessment
UDDI	Universal Description, Discovery and Integration
UI	User Interface
UML	Unified Modeling Language
UN/CEFACT	UN Centre for Trade Facilitation and E-business
URI	Uniform Resource Identifier
URL	Uniform Resource Locator
V2G	Vehicle-to-grid
VDE	Verband der Elektrotechnik Elektronik Informationstechnik e.V. (engl.: Association for Electrical, Electronic & Information Technologies)
VPP	Virtual Power Plant
W3C	World Wide Web Consortium
WD	Working Draft
WG	Working Group
WS	Web Service
WSDL	Web Service Definition Language
WXXM	Weather Exchange Model
XML	Extensible Markup Language
XSD	XML Schema Definitions

Part I
The CIM: Foundations

Chapter 1
Introduction

The Smart Grid is one of the dominating topics discussed today in the energy domain. Due to previous experiences as well as several national and international studies and roadmaps like [12], [108] or [92], it is generally accepted, that an appropriate Information and Communication Technologies (ICT) infrastructure is needed to control the future power type of power transmission and distribution grid and gather relevant data. Furthermore, the use of standards within this future infrastructure is indispensable as outlined by the aforementioned roadmaps and studies in order to reach a proper interoperability level.

A lot of new functions, services and use cases arise and stakeholders within the (electric) Smart Grid have to cope with those new challenges, which are, amongst others, highly focusing on interoperability of the components to be integrated. The IEC 61970/61968 Common Information Model (CIM) is undoubtedly one of the core standards of the future Smart Grid [108].

The CIM provides both a powerful data model (domain ontology) as well as various interface specifications and technology (communication and serialization) mappings. Its acceptance in the electric utilities sector and active standardization work is high and makes it one of the most established standards worldwide in the energy domain. From this book, you will get to know applicable knowledge about the CIM as well as its theoretical foundations, the history, the standardization processes and working groups and the corresponding technology behind the domain model.

1.1 Introduction of the Smart Grid

The term Smart Grid is one of the most frequently used words in the energy domain in the last few years. To establish a common basis, we provide the definition of Smart Grid for the context of this book in this section.

The term "Smart Grid" (an intelligent energy supply system) comprises the networking and control of intelligent generators, storage facilities, loads and network

M. Uslar et al.: The Common Information Model CIM, POWSYS, pp. 3–45.
springerlink.com

operating equipment in power transmission and distribution networks with the aid of Information and Communication Technologies. The objective is to ensure sustainable and environmentally sound power supply by means of transparent, energy - and cost-efficient, safe and reliable system operation.

The above definition is used in the German "Deutsche Kommission Elektrotechnik Elektronik Informationstechnik im DIN und VDE (DKE)" SMART.GRID mirror committee according to the International Electrotechnical Commission (IEC) Standardization Management Board Strategic Group 3 (SMB SG 3) "Smart Grid" and will be the very core of our definition like in [12]. Different power generation structures in various locations lead to the fact that a Smart Grid cannot be defined as a only one type and amount of attributes being possible. Regulation, different utility requirements, and, of course, natural resources are resulting in different requirements for the Smart Grid and the electric distribution and transmission grid. But there are not only widely differing requirements in the generation part of the grid, but also consumer structures have various characteristics in terms of energy prices or urban population density. Thus, the coordinated balance between power generation and consumption may be different in relation to the corresponding location. There are central aspects which can be defined as follows:

- It is necessary to integrate more variable generation of distributed generation and storage options.
- A holistic, intelligent energy supply system called called "Smart Grid" comprises active power distribution and transmission with new ICT-based technologies
- Distributed and decentralized grid with distributed coordination intelligence and data aggregation management is one of the core concepts
- Customer involvement in form of a so called prosumer (consumer and possible producer) residing in a smart building with electric vehicles and the necessary ICT-based equipment to support the decentralized decision making, bidirectionally communication with the utility, and, most of the time, smart-metering

Also, there are further international definitions which could be consulted to both define and coin the term Smart Grid like the ones following in the next paragraphs.

The European Technology Platform (ETP) "Smart Grids" [94] has defined a strategic vision for a so called "transition" to the Smart Grid. The use of common technical standards and protocols to achieve open access, interoperability and vendor independence of the components is one of the core demands. Also, bidirectional connection for both communication and electricity flow is considered as a key feature. From the viewpoint of the ETP "Smart Grids", the Smart Grid is defined as follows:

A Smart Grid is an electricity network that can intelligently integrate the actions of all users connected to it - generators, consumers - and those that do both - in order to efficiently deliver sustainable, economic and secure electricity supplies.

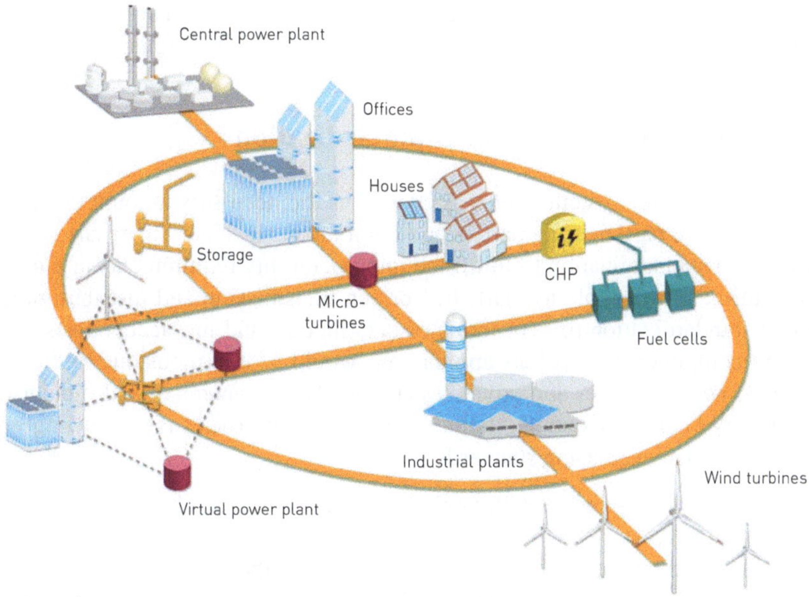

Fig. 1.1 The European Union's Smart Grid vision of the European Technology Platform SG [94]

The American National Institute of Standards and Technology (NIST)[1] has another definition as follows [19]:

The term "Smart Grid" refers to a modernization of the electricity delivery system so it monitors, protects and automatically optimizes the operation of its interconnected elements from the central and distributed generator through the high-voltage transmission network and the distribution system, to industrial users and building automation systems, to energy storage installations and to end-use consumers and their thermostats, electric vehicles, appliances and other household devices.

The definition shows that the NIST has a focus on the transition process to a Smart Grid. In general, every Smart Grid definition comprises technical aspects and domains as well as actors. As the definitions show, one key element of the Smart Grid transition is the integration of new technologies from the ICT domain with the existing grown infrastructure of automation and power distribution and transmission. As legacy processes and systems have to be integrated with new functionality and components, from ICT perspective, classic enterprise application integration (EAI) projects arise with all of their requirements and problems. Chapter 2 shows the basic concepts from the ICT-perspective on these issues. As systems are integrated, they have to agree on both a syntactical and semantical level

[1] http://www.nist.gov/smartgrid/index.cfm

on the concepts, profiles and serialization of the data to be exchanged between the systems. Therefore, a common information model is of highest importance.

1.2 General Motivation for Standardization

The requirement for system openness as well as the necessary amount of data exchange between participating parties and components inside a Smart Grid ecosystem leads to standardization to fulfill the interoperability requirements in a holistic architecture and to enable a smart, ICT-driven transmission and distribution grid. Without standardization (e.g. in terms of data models and interfaces) costs for integrating components as well as applications would be enormous due to the sheer number of new interfaces and processes arising. The German national standardization strategy [11], for example, defines amongst others issues the following goals which can be achieved using standards:

- Standardization can act as a strategic instrument to support both the economical and social success.
- Standardization relieves the regulatory activities of the government
- Standard Developing Organization (SDOs) foster technology convergence and advancement
- SDOs provide efficient processes and instruments for industrial participation and coordination

The various regional and international initiatives for Smart Grid standardization (see a summary from a project in [105] or [115][2]) outline the importance of standards in the Smart Grid domain. Most of them name and recommend single standards or families of standards, but they have one major recommendation in common - the IEC 61970/61968: CIM is seen as one of the core standards for the future Smart Grid.

The following sections and paragraphs are going to provide both a short overview on how Smart Grid standardization is seen in different regions of the world as well as the current status and its context and status for those roadmaps. The section focuses on the most prominent and relevant roadmaps as of time of writing the book, the companion web site[3] to the book will provide updates as well as additions and changes of these roadmaps.

1.3 Smart Grid Standardization Roadmaps and Their Focus on CIM

As the previous section has motivated, different roadmaps and strategies have evolved over the time in terms of both Smart grid functionality, utilization and standardization. This section of the book provides an overview on the most important

[2] Both were also basis for [12] and [5].

[3] www.61970.de

roadmap on regional and international level as well as a strong focus on the use of the CIM in this context. A large number of roadmaps directly name CIM as one of the core standards or at least one of the most important standards for the Smart Grid, some roadmaps focus on naming requirements, i.e. a common data model for data exchange, semantics and syntax for interfaces and pre-defined technology communication mappings.

1.3.1 The German Standardization Roadmap E-Energy / Smart Grid

In 2010, the German national Smart Grid (or E-Energy as the topic is often called in Germany) standardization roadmap [12] was published, originally initiated from the German Commission for Electrical, Electronic & Information Technologies. It focuses on Smart Grid's ICT and takes national and international standards into account. The roadmap considers existing initiatives and adopts and extends the recommendations for a German view. The development includes the collaboration of several experts from national standards developing organizations, governmental institutions, utilities as well as universities and research institutes.

After the identification and analysis of several studies and initiatives, a pre-study by OFFIS - Institute for Information Technology [97] was made available for the purpose of internal DKE discussion and later on for public comments and feedback of a congress and a commenting period. The German roadmap provides an overview on various international standardization efforts as well as specific recommendations for standards relating to different parts of the Smart Grid. Further, Smart Grids and standardization in general is introduced. Based on the analysis of several standardization roadmaps and related approaches, challenges in the use of standards and common factors were identified.

The identified key challenges consist of the coordination and integration of different industries, grid participants and grids itself, the harmonization of different models on data and semantic level, integration technologies, and of course, continuous security on the communication paths as well as stored data. The importance of international standards is beyond debate, but the regulatory, technical political and organizational aspects should be considered as well to create a sustainable Smart Grid.

In summary, the roadmap gives recommendations of standardization related aspects and issues. The 50 recommendations given are split into cross-cutting topics and domain specific areas as shown in table 1.1.

Apart from identifying core standards for the Smart Grid, especially in the ICT domain, the development of new or the improvement of existing standards is considered. With the constitution of a German Smart Grid Standardization steering committee in July 2010, the initiative has gained momentum and a continuous development of the roadmap is planned. It is planned that the future versions of the roadmap should include, among others, multi-utility aspects (like gas, heat and water), more communication standards (like market communication), High-Voltage

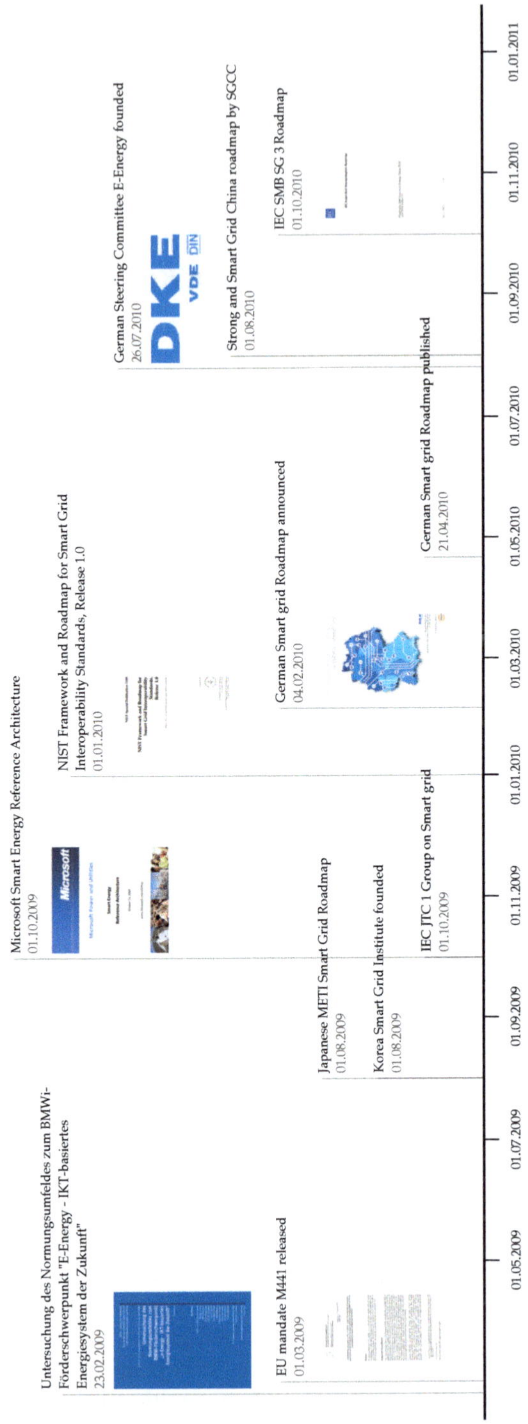

Fig. 1.2 Timeline and genealogy for SG roadmaps

Table 1.1 Scope of the German SG Roadmap

General recommendations
Regulatory and legislative recommendations
Recommendations on information security, privacy and data protection
Recommendations on the area of communications
Recommendations on the areas of architectures, communication and power system management processes
Recommendations on the safety, reliability and durability of products: Domain-specific areas
Recommendations for the area of Active Distribution Systems
Recommendations for the area of Smart Meters
Recommendations on the area of Electromobility
Recommendations for the area of Storage
Recommendations for the area of Load Management/Demand Response
Recommendations for the area of Building and Inhouse Automation
Recommendations for the area of Distributed Generation
Recommendations for the area of Transmission Systems

Direct Current (HVDC) as well as Flexible Alternating Current Transmission System (FACTS).

The focus on CIM within this roadmap is as follows. In German standardization there is no matching mirror committee for the IEC Technical Committee(TC) 57 Working Group (WG) 13, 14, 16 and 19, so the issues for German CIM requirements are dealt with in the DKE K 952 which is the mirror for IEC TC 57 in general. Therefore, the issues are dealt holistic with. In general, the German roadmap recommends the standards in table 1.2. As one can see, CIM is recommended directly but there are further detailed recommendation in the roadmap. Regulatory changes are called for with recommendation SG-RE 1 which suggest for using CIM and Extensible Markup Language (XML) for future market communications. SG-K-3 calls for using the Seamless Integration Reference Architecture (SIA) alongside the CIM (see chapter 1.6.1, SG-K-4 for extending the CIM to a multi-utility model. SG-ANLT-2 deals with the recommendation to integrate CIM and IEC 61850. The recommendation SG-ANLT-4 calls for a repository for both use cases and Unified Modeling Language (UML) models like CIM. SG-AV-1 directly recommends using the CIM for interfaces at Distribution Management System (DMS) level. Pricing and tariffing models using CIM are requested in SG-EM-3. Within the whole roadmap, several recommendations directly address the CIM and its application which highlights its importance. The roadmap is under constant review and a first update report was provided in January 2011 [13].

Table 1.2 Recommended standards within the German Smart Grid Roadmap

IEC 62357	Seamless Integration Reference Architecture
IEC 62351	Security for Smart Grid
IEC 61968/61970	Common Information model (CIM)
IEC 61850, 61850-7-4XX	Substation Automation Systems (SAS) Communications, Decentralized Energy Resources(DER)
IEC 60870	Transport protocols
IEC 61400	Communications for monitoring and control of wind power plants
IEC 61334	Device Language Message specification (DLMS)
IEC 62056	COmpanion Specification for Energy Metering (COSEM)
IEC 62325	Market Communications Using CIM

1.3.2 NIST Framework and Roadmap for Smart Grid Interoperability Standards, Version 1.0

The Department of Commerce (DoC) of the United States was authorized by the Energy Independence and Security Act (EISA) in 2007 to coordinate the development of an interoperability framework. The NIST was put in charge to coordinate the efforts to achieve interoperability in the Smart Grid system and treats the standards in the ICT domain including protocols and data models with special respect. One of the first actions to promote the efforts was to propose a three-phase plan to ensure the continuous development and implementation of standards. This plan consists of the following action items:

Engagement of stakeholders via a participatory public process: Identifying applicable standards and requirements but also gaps in currently available standards. For this purpose, NIST has compiled and incorporated stakeholder inputs from several public workshops with the support of outside technical experts. In addition, technical contributions from technical working groups and a cyber-security coordination task group were also taken into account.

Establishing a Smart Grid Interoperability Panel: For ensuring continued development of interoperability standards, an organizational forum was launched in November 2009, the so called Smart Grid Interoperability Panel (SGIP).

Development and implementation of a framework for conformity testing and certification: According to NIST, testing and certification how standards are implemented in Smart Grid devices, systems and processes is a very core issue to ensure interoperability, interchangeability and both safety and security.

Further, NIST designed a priority action plan (PAP) to support and foster the development of important standards, the first PAP version comprises 18 tasks.

In 2009, the first version of the overall envisioned framework was published [19]. It includes an abstract conceptual model consisting of about 80 standards which are related to the Smart Grid. This relation can either be directly or on a meta-level. Additionally, 15 key areas and gaps have been identified. These gaps have to be closed with either new or improved existing standards. Furthermore, 16 standards for the future are recommended by this document, which can be seen as core standards for the future Smart Grid.

The latter document "NIST Framework and Roadmap for Smart Grid Interoperability Standards, Release 1.0" [92], can be seen as the output of the first phase from the earlier introduced NIST plan. Despite the fact that this document seems well advances, the standards listed should not be regard as neither exhaustive nor exclusionary according to the roadmap itself. Besides the already mentioned sixteen core standards, many additional standards, guidelines, specifications, and reports are presented. Additionally to the NIST Framework and Roadmap [92], a document with further focus on Smart Grid Cyber Security was published [91].

To monitor and represent the progress of the Smart Grid activities the NIST provides the Smart Grid Collaboration Site [93]. As time of writing this chapter, version 2 of the roadmap is under development and likely to be rolled out soon after publishing this book. CIM has been named one of the sixteen core standards to be rolled out in future US Smart Grid and NIST aims at extending the model (for DER, Electrical Vehicle (EV), and things like Weather data) with its PAP items for better suitability to North American Smart grid needs.

1.3.3 IEC SMB Strategy Group 3 Report

In 2010, a draft report was published by the SG 3 (Strategic Group 'Smart Grids'), which was appointed by the IEC Standardization Management Board (SMB) to develop a Smart Grid roadmap. This report holds mainly IEC Standards as well as high level recommendations to use these standards. Later in 2010, based on this draft, the IEC Smart Grid Standardization Roadmap Edition 1.0 [108] was released.

One of the main topics is to achieve a high level of syntactic and semantic interoperability between involved components and solutions and to improve the monitoring and control functions of these components. This aspect is seen crucial to achieve the desired Smart Grid.

The following points are seen as main drivers for the required changing process:

- Increasing energy consumption
- Further spread of DER
- Competitive market prices
- Maturing infrastructure
- Sustainability of generation and distribution
- Security of energy supply

The IEC roadmap focuses on identifying existing standards in the IEC portfolio, tries to find potential gaps between those standards, but does not analyze standards from other standards developing organizations. IEC relied on the expert's description of the major applications of the Smart Grid to identify the necessary requirements. A top-down approach was used for identifying standards as well as gaps between them.

Based on the identified requirements, tasks and necessities for standardization were derived by the IEC. The procedure consisted of three main tasks:

- Capturing requirements
- Evaluating and rating of the impact of those requirements
- Assessing the requirements to functions which have architectural impact

Within this procedure, sub-systems were described in the same structure of five views, namely description, requirements, existing standards, gaps and recommendations. As a result, more than 100 standards and TCs have been identified, described and prioritized. Out of these standards, five standard series are ranked as 'core' and nine are ranked as 'high priority' standards, based on differentiated analysis. Furthermore, the roadmap includes thirteen main application areas, six general topics and more than 40 recommendations overall [108].

In the document, SG3 recommends to promote the already existing IEC standards in the Smart Grid domain. Also, a close cooperation with the NIST roadmap is highly suggested. Additionally, the IEC Mapping Tool was developed. It provides an overview on Smart Grid standards and adds several filter criteria for an effective selection and usage of existing standards. It is available on the IEC Smart Grid website (see [25]). Latest information regarding Smart Grid activities of the IEC can be accessed as well at the IEC Smart Grid site, see [24].

CIM has been named several times in the recommendations of the roadmap. As the SIA IEC 62357 (see chapter 1.6.1) is in the core of the IEC roadmap (p. 24). the discussion around Service-oriented architecture (SOA) and data models is based on both CIM and 61850. All six major communication recommendations deal with the use of CIM. CIM should be the basic data model for SOA-based Enterprise Service Bus (ESB) implementations and architectures, provide proper profiles for easier integration, be future-proof as it can be extended, e.g. to DLMS/COSEM and Advanced Metering Infrastructure (AMI) and harmonized alongside IEC 61850 to provide seamless vertical integration and end-to-end security. In terms of distribution management, the IEC 61968 series should be further developed and promoted. Normative messages and mappings to MultiSpeak (R) should be developed, interoperability testing extended and AMI and home area network (HAN) models incorporated. One important issue is to extend the CIM to DER and Fleet generation modeling. IEC 62326 should be promoted for market communications, Vehicle-to-grid (V2G) communications and pricing models.

1.3.4 German E-Energy Projects Study on Standardization

In cooperation with the Federal Ministry of Environment, Nature Conservation and Nuclear Safety (BMU), the Federal Ministry of Economics and Technology (BMWi) funded the German E-Energy program, which has been established in

2008. Spread across Germany, a total of six model regions are carrying out demonstration projects with different focuses. One main target is to drive on the so called "Internet of Energy" topic. The BMWi commissioned the study to analyze the standardization environment of all six model regions to identify the common basis in this domain. The intention was to consistently provide information about existing standards, standards in process of creation or renewing as well as information about additional standards for ICT based energy systems and their use in the specific model region and its projects.

To gain the necessary information for this study, the standardization environment was examined based on expert interviews, literature research and the former experiences of the project consortium in the standardization participation. A standards overview was created, which categorizes the identified standards into vertical and horizontal standards. In addition, the initiatives of other countries and their experiences in this domain are documented and demonstrated how to adapt this additional information for Germany.

The study gives recommendations on different domains of standards like:

- Software engineering
- Software architecture concepts
- Functional standards architecture
- CIM
- Substation automation
- DER communication
- Tele-control
- Programmable logic controllers
- Market Communication
- Security
- Home automation
- Smart Metering

One main recommendation concerns the IEC TC 57 Reference Architecture (IEC 62357) and, thus, of course the CIM series of standards (see chapter 1.6.1). Another main point was the use of the IEC 61850 regarding substation automation, communication as well as tele-control of DER. Additionally, the cross-cutting standards like the security standards IEC 62351 and IEC 62443 were considered. International and national best practices and the requirements of the German Association of Energy and Water Industries (BDEW) had further influence on the recommendations regarding security.

To be useful for all Smart Grid projects initiated, most recommendations are as unspecific as possible. Besides the comprehensive IEC standards, the study focuses mainly on converging technologies between the ICT and the automation level domains. The resulting document was also used as input for the national German roadmap [12] in chapter 1.3.1 as well as an input for the IEC roadmap [108], further described in chapter 1.3.3. CIM is highly recommended with the very same drivers and requirements as in the German roadmap which was published later on as well as in the context of additional standards like the IEC 62559 for Use Cases from the IntelliGrid roadmap.

1.3.5 Microsoft Smart Energy Reference Architecture

In 2009, Microsoft Corporation published the Smart Energy Reference Architecture (SERA) for the envisioned smart energy ecosystem. This work defines a reference architecture as a consistent framework to guide implementations within the smart energy domain. It identifies technological aspects and solution concepts from the electricity value chain from the end-user up to the generation.

Microsoft intends to accelerate solutions development and to support more cost-effective, secure and scalable implementations. The recommended solutions consist mainly of Microsoft products in the Smart grid context. To take into account the power utilities IT infrastructure, a close cooperation with power industry partners has been carried out. The best practices extracted from this collaboration have been included in the reference architecture.

SERA appreciates the NIST work and wants to be seen as a bridge from NIST standards to specific Microsoft products and technologies. The architecture itself describes characteristics and requirements of implementations and is split into the following seven components:

- Approach
- User Experience
- Collaboration
- Information
- Integration
- Application Architecture
- Security

The reference architecture includes the description of quality attributes like performance and interoperability, interfaces to provide users with access to role-based information and services, and potential information and data exchange styles for collaboration. Approaches concerning representation, management and exchange of Smart Grid related information are outlined and the integrating and participation of systems and data are proposed with SOA supported cloud services. For the domains application architecture and security, the requirements are briefly summarized.

The IEC CIM is one of the mentioned and adapted standards into the SERA. The technology independent standards should enable software vendors to create interoperable solutions. As CIM is seen in the very context with the adoption of the NIST roadmap by Microsoft, it is also identified as one of the basic blocks for the future Smart Grid. Being an ICT vendor, CIM is in the very core as the data model for the data integration and enterprise service-bus strategy of Microsoft. Integration and message exchange based on XML and SOAP can be easily conducted using CIM.

1.3.6 CIGRE D2.24 - EMS Architecture for the 21st Century

The WG D2.24 of the Conseil International des Grand Reseaux Electriques (CIGRE) focuses on 'Energy management system (EMS) Architectures for the 21st Century' and wants to produce a complete and consistent specification of such an EMS-architecture [8]. Furthermore, it is intended to gain wide adoption in practice and to facilitate its implementation as a de-facto standard. To reach this goal, D2.24 wants to create a common architecture vision and thus identify the common requirements for the next generation energy and market management systems. The key focus is thereby to achieve a highly interoperable, modular and re-usable architecture.

A set of common requirements has been identified and the WG describes the architecture and the standard components. In the next generation energy and market management systems architecture, the focus lies on real-time coupled system between transmission grids and market systems. This should also be extendable to generation and distribution as well. Concerning the architecture development, ten design principles were defined and recommended, which are as follows [8]:

- Component-based and service-oriented architecture to enable integration and reusability within the utility
- Modularity based on business processes components
- High speed bus for real-time message exchange
- Common information model for data standards built on industry standards
- Common user interface
- Security layer
- Common modeling tools integrating display and data maintenance transportable to future technologies and scalable for future use cases

The CIGRE WG D2.24 is not responsible for setting any standards, but only for setting up a vision and its requirements. A liaison between IEC TC 57 and CIGRE D2 has been established to incorporate IEC standards into the architecture. Therefore, the recommended architecture will mainly consist of IEC TC 57 standards.

As the CIGRE D2.24 strongly focused on EMS; the CIM being an EMS application programming interface (API) was in the very focus, also the ideas of implementing SOA for Supervisory Control and Data Acquisition (SCADA) and EMS/DMS. Still, no clear recommendations for standards were ever directly stated. The current work status of the group is unclear, it seems as if there are no current activities.

1.3.7 EU Mandates to CEN/CENELEC/ETSI - SG-CG and SM-CG

The European Union promotes advanced metering systems through different regulatory measures. This should lead to a wide European roll-out of smart meters. The main goal is to increase energy savings by developing a greater awareness of energy consumption at the end user. To push this topic, the European Union issued the

EU Mandate M/441, which concerns the standardization of smart meter functionalities and communications interfaces for the use in Europe. The standards should be seen as voluntary technical specifications and technical rules for the products and systems in this context.

The Smart Meters Co-Ordination Group (SM-CG) was formed by European Committee for Standardization (CEN) , European Committee for Electrotechnical Standardization (CENELEC) and European Telecommunications Standards Institute (ETSI) to consider the main issues addressed by the mandate. The following six aspects of smart metering are considered and the prevailing standards were examined:

- Reading and transmission of measurements
- Two-way communication between the meter and a market participant (biller)
- Support by the meter for various tariff models and payment systems
- Remote meter deactivation and start/finish of supply
- Communication with devices in the household
- Support of a display or interface in the household for display of the meter data in real time

Decisions about functionalities should be left to the individual countries. So this list should not be considered to be a minimum list of necessary functionalities and, above all, it is not proposed to establish a minimum set through standards. Six functionalities and the associated responsibilities as well as already existing standards are delegated to the individual standardization bodies within the SM-CG (see figure 1.3.

The European Standardization Organizations (ESO) are responsible fore accepting new work item proposals to ensure that new technologies, protocols and procedures can be adopted from the market. But before creating new standards, the improvement of existing standards shall be preferred. As of now there is no detailed timetable for the future process.

The mandate M/490 is an additional mandate similar to M/441 to the three European standardization bodies CEN/CENELEC and ETSI. The Joint Working Group "Smart Grid" (JWG-SG) created a report which consists of recommendations for the following topics:

- Terminology
- System Aspects
- Data communication reference architecture
- Data communication Interfaces with the focus on sectional standards like e.g. DMS, SCADA, Data models, Enterprise Resource Planning (ERP) interfaces
- Security

The scope of the mandate is defined as following (see also [83] for an extended version):

"The objective of this mandate is to develop or update a set of consistent standards within a common European framework that integrating a variety of digital computing and communication technologies and electrical architectures, and associated processes and services, that will achieve interoperability and will enable or

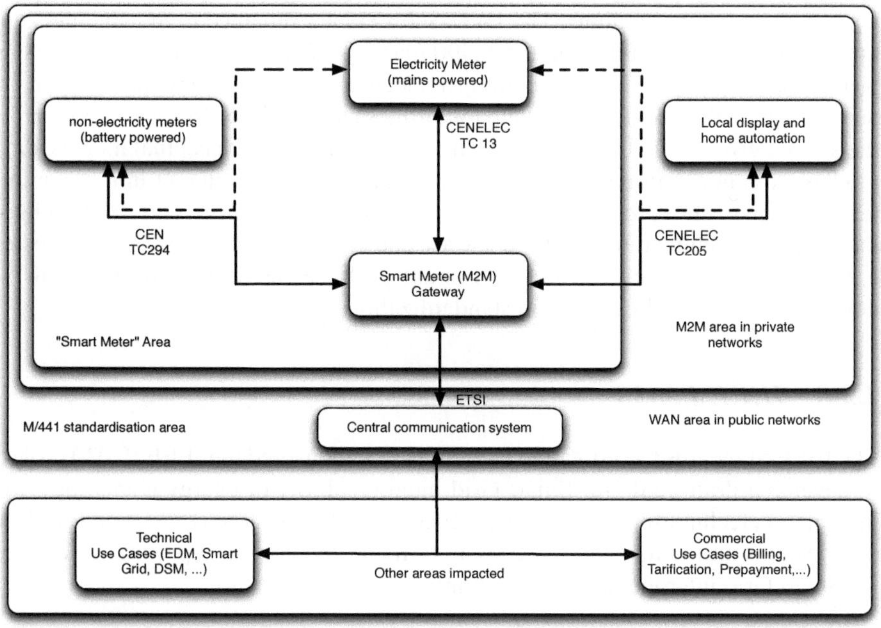

Fig. 1.3 M/441 overview

facilitate the implementation in Europe of the different high level Smart Grid services and functionalities as defined by the Smart Grid Task Force that will be flexible enough to accommodate future developments. Building, Industry, Appliances and Home automation are out of the scope of this mandate; however, their interfaces with the Smart Grid and related services have to be treated under this mandate.

It will answer the technical and organizational needs for sustainable state of the art Smart Grid Information Security (SGIS), Data protection and privacy (DPP), enabling the collection, utilization, processing, storage, transmission and erasure of all information to be protected for all participating actors. This will enable smart grid services through a Smart Grid information and communication system that is inherently secure by design within the critical infrastructure of transmission and distribution networks, as well as within the connected properties (buildings, charging station up to the final nodes). This should be done in a way that is compatible with all relevant legal requirements, i.e. consumer data protection and privacy rights, metrology and daily business operations, and that is ensuring that rights of all consumers, including the vulnerable ones, are protected."

This mandated work will be co-ordinated with the outcomes of the existing mandates M/441 and M/468 to ensure a coherent standardization framework (a bi-directional process). Because of the huge number of stakeholders, the necessary speed, the many international activities and the still changing recommended solutions make standardization a difficult task for the ESO. The JWG report investigated the status of European standardization but should not duplicate the extensive work already done in other regions. The JWG-SG was later renamed to SG-CG and kept up with their work on fulfilling the M490 mandate. Therefore, four working teams were established. One aims at providing a selection of standards and their gaps in order to improve them and provide help for utilities and is called First Set of Standards, the second group called Reference Architecture deals with conceptual models, data models, communication models and how standards have to interact in those architectural views, the security group addresses smart grid security and safety within the reference architecture and the Use Case and sustainable processes group creates a meaningful repository of European Smart grid use cases and processes how to develop them and address them in standardization.

1.3.8 IEEE P2030

The IEEE Standards Coordinating Committee 21 sponsored the IEEE P2030 project to develop a draft guide for Smart Grid interoperability of energy technology and information technology operation with Electric Power System (EPS), and End-Use Applications and Loads [77]. According to IEEE, the goal is to provide guidelines for Smart Grid interoperability including:

- Terminology
- Characteristics
- Functional performance
- Evaluation criteria
- Application of engineering principles for Smart Grid interoperability

The EISA from 2007 was the main driver for this project, so a close cooperation with NIST is to be expected. As a result, the recommendations will be highly similar and the recommended standards are overlapping.

The large number of participants were divided into three main task forces (Power Engineering Technology, Information technology and Communications) and an unprecedented collaboration of the three IEEE Societies will be expected. Hereby, the Power Engineering Technology section will focus on the functional requirements of interoperability. The sections Information Technology and Communication will seize on these identified requirements and will interoperability guidelines for protocols, architecture, transport, application, security, privacy and other. CIM is (of course) not a direct focus of IEEE, but the recommendation of a common data model is given.

1.3.9 Japanese METI International Standardization Roadmap for Smart Grid

A strategy group was founded in August 2009 with the aim of promoting Japanese activities in international standardization [81], because of an Initiative started by the Ministry of Economy, Trade and Industry (METI). The strategy group represents the opinion that the desired flexibility and expandability of the Smart Grid can only be achieved with an appropriate degree of standardization, which is a fundamental element of interoperability.

The first report of this roadmap was released in 2010. It provides a roadmap of standardization efforts generated in close cooperation with other international standardization organizations. The following seven main fields were identified:

- Wide-Area Awareness in Transmission
- Supply Side Energy Storage
- Distribution Grid Management
- Demand/Response
- Demand-Side Energy Storage
- Electric Vehicles

In addition to those seven fields, special core aspects for the Japanese economy were identified. Because of the close cooperation with international standardization organizations, the resulting recommendations are nearly fully congruent with the other international roadmap recommendations. In the near future, a further collaboration with other countries is planned as well as an examination of the need for an implementation body by a private sector consortium along with government assistance. As a close cooperation with IEC was in the Japanese focus, their recommendations regarding the CIM do not differ.

1.3.10 SGCC - Strong and Smart Grid China

The State Grid Corporation of China (SGCC) Framework and Roadmap for Strong and Smart Grid Standards [107] was published in 2009 as a national guideline to develop corporate standards. Additionally, it is used as a reference for industrial, national and international standards. The SGCC itself is a state-owned company which operates the most part of Chinas power grid.

This framework should have a huge impact on all Smart Grid product vendors because of the increasing demand and the Chinese market size. The main goal of this framework is to support the creation of an intelligent power system encompassing power generation, transmission, transformation, distribution, consumption and dispatching. The Chinese Smart Grid is planned to become a more integrated and intelligent platform for the Internet of things, Internet network and communication network. Additionally, radio and TV networks should be integrated beyond the power grid function as a carrier of transmission and distribution of electricity. On creating the roadmap, different existing approaches were considered. Among others, the work of the IEC SG 3, NIST Interoperability Roadmap, IEEE P2030, CEN/CENELEC/ETSI Working Groups, German Standardization Roadmap and the Japanese METI Roadmap were taken into account.

The first released version hierarchically divides its scope into eight domains, 26 technical fields and 92 series or parts of standards. Existing standards and gaps were identified in the following eight domains, cover planning, power generation, transmission, substation equipment and communication, distribution, utilization, dispatching and ICT. The authors examined domestic and international Smart Grid relevant standards. The result includes 22 core standards which have a close relation to Smart Grid construction and may be able to support the framework. The set of this standard can be divided into ten domestic ones and twelve international standards which are listed as follows:

- ISO/IEC PAS 62559: Terminology and modeling of Smart Grid
- IEC 61850: Standard series on Substation Communication network and System
- IEC 61968: Interface of Power Company Data Exchange Platform - Distribution Management System
- OGC Open GIS: Specifications on Open Geographical Data Interoperability
- IEEE 1547: Technology Regulations on Integration of Distributed Generations into Power Grid
- IEC 61851: Standard Series on Electric Vehicle Charging and Discharging
- IEC 61970: Standard Series on Application Program Interface of Energy Management Systems (EMS)
- IEC 60870: Standard Series on Transmission Control Protocol
- IEC 62351: Power System Management and Associated Information Exchange - Data and Communications Security
- IEC 62357: Power System Control and Associated Communications - Object Model, Service Facilities and Protocol Architecture with Reference
- ISO/IEC 27000 Series: Standard Series on Information Security Management System
- ISO/IEC 15408: Information Technology Security Evaluation Criteria

The plan for the future encompasses the three phases Planning and Trial, All-around construction and Leading and Enhancing from 2009 to 2010, 2011 to 2015 and, respectively, 2016 to 2020 including the long-term goal to further develop and improve the SGCC framework and roadmap and to participate in the formulation of national and international standards. Until the release of the roadmap, SGCC already started 228 demonstration projects of 21 categories, spread over 26 provinces and municipalities since August 2009 to realize their Smart Grid.

With the special regard to CIM, the roadmap outlines that CIM is an open, accepted and widely adopted standards in terms of China and national extensions like the CIM/E, CIM/G or CIM/S formats exist which will be brought into international standardization at IEC TC 57 level.

1.3.11 Korea's Smart Grid Roadmap

A roadmap to build a nationwide smart grid until 2030 was designed by the Korean Smart Grid Institute under the Ministry of Knowledge Economy (MKE). To reach this goal, a three phase plan was created with focus on primarily five domains. These domains include Smart Power Grid, Smart Consumer, Smart Transportation, Smart Renewables and Smart Electricity Service to enable the Smart Grid.

The first phase was appointed between 2010 and 2012 and focuses on establishing a proper technical platform based on the results generated by ten already ongoing demonstrations projects which are started in 2005. These projects concentrate on the following tasks:

- Development of Korean Energy Management System
- Development of Intelligent Transmission Network Monitoring and Operation System
- IT Based Control System for Bulk Power Transmission
- Development of Prototype for Advanced Substation Automation System based on the Digital Control Technology
- The Development of Power Equipment Monitoring System using Active Telemetrics
- Development of Intelligent Distribution Management System
- Development of Power Line Communication (PLC) Ubiquitous Technology
- Power Semiconductor for Dispersed Generation and Industrial Inverter Application
- Development of integration EMS for the micro grid and application technology to real site
- Consumer Portal System for IT-Based Energy Services Business

The second phase planned from 2013 to 2020 aims to provide intelligent services including Virtual Power Plant (VPP) infrastructure, Smart power trading Systems, Smart new renewables-linked technologies, AMI, V2G, and a Urban Smart Power Grid development on a wide area scale.

The third and last phase from 2021-2030 aims to establish the results on a nationwide scale with a national Smart Power Grid operation, AMI-based automatic power trading, EV and charging service generalization, large scale new renewable supply infrastructure development as well as an integrated power trading including service development.

Until now, there has been no specific roadmap in using international standards in the Korean Smart Grid, but cooperations with home appliances projects like the German E-Energy projects especially the German Kellendonk EEbus home automation wrapper series which will be ZigBee SEP or CIM based.

1.3.12 ISO/IEC JTC 1 Special Working Group on Smart Grid (SWG-Smart Grid)

The Joint Technical Committee 1(JTC1) recognized the continuing and important evolution of Smart Grid technologies and established the SWG-Smart Grid on a meeting in October 2009 in Tel Aviv [1]. Many standards consortia all over the world are planning to develop Smart Grid standards, so the JTC 1 believes that it has specific interest in this area on a continuing basis and wants to participate.

The terms of reference for the SWG include the following points [1]:

- To identify market requirements and standardization gaps for the Smart Grid especially interoperability standards in an international context
- To encourage the JTC 1 Subcommittees (SCs) to address the need for standardization of the identified points

- To promote the developed International Standards
- To coordinate JTC 1 Smart Grid programs of work with other Organizations
- To develop JTC 1 Smart Grid strategic plan recommendations

The primary focus lies on international standards supporting interoperability of Smart electric Grid technology. The extension needed to include gas, water and other metering systems and interfaces with home and building automation systems that are dependent on the same Smart electric Grid information and communications infrastructure are also part of the focus. The national bodies of Canada, France, Germany, Japan, Republic of Korea, The Netherlands, Singapore, United Kingdom, The United States and also the JTC 1 Subcommittees SC 6, SC 22, SC 25, SC 32, and WG7 have nominated experts especially for the SWG-Smart Grid. Liaisons and relationships with groups like IEC SMB SG 3 have also been requested, recently.

The group accepted to use the IEC SMB SG 3 architectural framework as a starting point as the first agreements and to focus on interoperability layer categories 1 through 7 (Basic Connectivity through Business Objectives). Unfortunately, an explicit roadmap has not been published until now.

1.3.13 Summary: Recommended Core Standards on Worldwide Level

Different countries, organizations and vendors came up with their own roadmaps regarding the Smart Grid standardization. Some countries focus on solutions at the customer's site which often have different focus from home appliances for demand response (DR), peak shaping, home automation, other countries mainly focus on reducing non-technical losses through smart metering or improving the outage management for radial feeder systems.

One main scope overall seems to be to provide the reliability of supply for the digital economy, focusing on markets and economic benefits for the country. Another strong point is to integrate more sustainable energy, trying to reduce carbon dioxide emissions and to cope with distributed, renewable generation like Micro Combined Heat and Power (CHP), photovoltaics, and fuel cells as well as with electric vehicles. A one-size-fits-all solution for those Smart Grid requirements both in terms of technical solutions and its corresponding standardization requirements is unlikely to be found.

In Fig. 1.4, an overview about the approaches currently considering or making recommendations for standards are listed, therefore IEEE P2030, Japan's, Korea's and the JTC1's roadmap are missing. Planned extension of standards by the SGCC Framework are high-lighted in light-grey.

The main intention of providing this table is to facilitate an overview of international standards like IEC or ISO standards, so only few national standards have been integrated. Looking through the table, the most striking and common core standards are from the IEC TC 57 which will be briefly introduced in the following section of this book. According to the number of recommendations, the following TC 57 standards are of highest importance from the perspective of most experts: Since national perspectives differ, organizational standards e.g. like IEEE ones have less impact on worldwide scale.

- IEC TR 62357: Reference Architecture
- IEC 61968/61970: Common Information Model for EMS and DMS
- IEC 61850: Intelligent Electronic Device (IED) Communications at Substation level and DER
- IEC 62351: Vertical security for the TR 62357
- IEC 60870: Telecontrol protocols
- IEC 62541: OPC UA - OPC Unified Architecture, Automation Standard
- IEC 62325: Market Communications using CIM

Within chapter 1.6 of this book, we provide a short overview on those standards.

1.4 History and Motivation for Development of CIM

The CIM, originally designed as an EMS-API, the CIM was developed being a internal database model (relational schema) for EMS and SCADA system but soon changed into a much more useful object-oriented approach to model all the relevant objects and their relations in the scope of electric distribution, transmission and generation. The domain ontology is pretty large, covering most aspects being focused on subparts of the model like generation, outage, documentation, transmission, wires, measurements and so on. Major use cases for the CIM are system integration using pre-defined interfaces between the DMS IT and automation parts, custom system integration using XML based payloads for semantically sound coupling of systems and serializing topology data using Resource Description Framework (RDF) for state estimation, SCADA and geographic information system (GIS) exchange. Most of IT Enterprise Application Integration (EAI) and SOA aspects in a modern electric utility can be covered using the CIM. A Gartner survey [20], carried out in 2010, outlines the widespread application of the CIM in the electric utility domain and pilot-projects and systems.

The Common Information Model was originally created by the Electric Power Research Institute (EPRI) in the mid 90s at the Control Center Application Programming Interface (CCAPI) project. The CIM was originally designed to solve the problem of vendor lock-in created by the EMS vendors. Over time, it became apparent that having a common definition of the data being passed between applications was a critical component. The concept of having a common information model was determined to be a fundamental problem that needed to be solved and this was the starting point of the CIM. Over the time, the CIM has slowly out-grown its original purpose.

The efforts on common data modeling and interfacing progressed and the next main target was to promote the CIM not just to utilities, but to vendors. The chosen way to do this was to turn the CIM in an international standard. The IEC started its efforts to establish the CIM and chartered the TC 57: Power System Management

IEC Standard		Approach								
Standard	Description	DKE Normungs-roadmap Smart Grid 1.0	NIST IOP Framework Roadmap 1.0	SMB SG 3 IEC	BMWi E-Energy Program Recommendations OFFIS	BDI - Internet der Energie	Microsoft SERA	CIGRE D2.24	Japan's Roadmap to international standardization for Smart Grid (preliminary)	SGCC Framework
IEC 60870	Communications protocol	■			■			■		■
IEC 60870-5	Telecontrol, EMS, DMS, DA, SA	■		High				■		
IEC 60870-6 / TASE.2	Inter-control center communications TASE.2 Inter Control Center Communication EMS, DMS	■	■	High			■	■		
IEC 61334	DLMS				■	■		■	■	
IEC 61400-25	Wind Power Communication EMS, DMS, DER	■		High	■	■		■		
IEC 61499	SPS and Automation, Profile for IEC 61850				■					
IEC 61850	Substation automation and protection, Distributed generation, Wind parks, Hydro, E-Mobility	■	■	■	■	■	■	■	■	■
IEC 61850-7-410	Hydro Energy Communication EMS, DMS, DA, SA, DER	■		High				■		
IEC 61850-7-420	Distributed Energy Communication DMS, DA, SA, DER, EMS	■		High	■	■		■	■	
IEC 61851	EV-Communication Smart Home, e-Mobility	■		High						■
IEC 61968	Distribution Management, System Interfaces for Distribution Management Systems, DCIM (CIM for Distribution)	■		■		■		■	■	■
IEC 61968/61970	Application level energy management system interfaces, CIM (Common Information Model), Domain Ontology, Interface, Exchange format, Profile, Process blueprints, CIM (Common Information Model) EMS, DMS, DA, SA, DER, AMI, DR, E-Storage	■	■	■	■	■	■	■	■	simplyfied CIM approach
IEC 61970	Energy Management, Application level energy management system interfaces, Core CIM	■		■		■		■	■	■
IEC 62051-54/58-59	Metering Standards DMS, DER, AMI, DR, Smart Home, E-Storage, E-Mobility	■		High						
IEC 62056	COSEM DMS, DER, AMI, DR, Smart Home, E-Storage, E-Mobility			High	■	■			■	
IEC 62325	Market communications using CIM	■			■			■		
IEC 62351	SIA Security	■	■	■	■	■	■	■	■	■
IEC 62351 Parts 1-8	Information security for power system control operations	■	■	■	■	■	■	■		
IEC 62357	IEC 62357 Reference Architecture – Service-oriented Architecture, EMS, DMS, Metering, Security, Energy Management Systems, Distribution management Systems	■	■	■	■			■		■
IEC 62443	General Security					■				
IEC 62541	OPC UA (Automation))	■					■	■		
IEC TR 61334	DLMS, Distribution Line Message Service			High			■			

Legend

High: standard considered of high relevance (only used within IEC SMB 3)

■ (black): addressed standard (core standard within IEC SMB 3)

simplyfied CIM approach: extension of the CIM (only used within SGCC)

Fig. 1.4 Overview on Roadmaps and Standards

and Associated Information Exchange. The TC 57 initially started with submitting the following drafts for standardization:

- Common Information Model (CIM)
- Generic Interface Definition (GID)
- Common Power System Model (CPSM)

The next step was to create working groups to add additional pieces to the information model. Each sub-area was assigned to small teams of interested parties. The current existing relevant working groups with focus on CIM are:

- WG 13 - Energy Management System Application Program Interface(EMS - API)
- WG 14 - System Interfaces for Distribution Management (SIDM)
- WG 16 - Deregulated Energy Market Communications
- WG 19 - Long Term Interoperability within IEC TC 57 Working Groups

These working groups design a series standards which are then assigned a standard number from the IEC. Each standard may include different parts, which are also numbered as a sub standard. With those, there are now parts of the standard which covers additional implementations and applications. Among those families of standards, the IEC 61970 and IEC 61968 are the most important ones.

The IEC TC 57 is focused on defining and revising the official standard within their SDO rules, so another forum was created to provide a venue for utilities, vendors, consultants and integrators who use the CIM. This forum is called the CIM users group (CIMug). The CIMug hosts shared information about the CIM and offers also a channel to the standards organization to make suggestions. Chapter 7 of this book deals with further information on the CIMug and how to participate in this group.

In the meantime, the EPRI takes a continued role in the CIM environment and conduct continuous research on areas where the CIM needs additional definitions or visibility. Additionally, EPRI coordinates the annual interoperability testing and tries to introduce CIM to other research activities where it can be used in larger initiatives [18].

1.5 Main Applications of the CIM

Outgrowing its original purpose, the CIM has become more than the envisioned data model and interface definition. Since the original development, not only the overall scope and modeling paradigm has changed, but also the core applications have been extended. There are different main use cases of the CIM nowadays.

First main use case of the CIM is based on using the CIM as a large domain ontology providing a vocabulary to express business messages which are exchanged between various system intra-utility. Previously defined as just an interface standard, the semantics to be exchanged between systems both intra- and inter utility have to be defined. Therefore, a Common Information model having common semantics is of interest. Alongside this, rules on how to design payloads and serializations are needed. So it is possible to do a custom-integration with highly standardized semantics and syntax of the business objects exchanged making for a strong coverage of standardized CIM objects and still maintaining the flexibility of the process data's structure being exchanged. Furthermore, as CIM is defined as an UML model, model-driven development and code generation for ECore, Eclipse RCP with EMF and GMF are a logical step. Using object-persistency stores, CIM can also be used for internal business object storage and representation.

Where the first main use case emphasizes the importance of the interfaces, the data exchanged is in the focus of main application two - the exchange of topology data using CIM. For both transmission and distribution grid, profiles (so called subsets of the CIM defining (choosing) objects and relations between them used in the very specific context) are defined to serialize the graph of the power grid in CIM/XML which is a W3C RDF-based serialization.

Using graph-like data structures, serialization, checking and making difference files for topology changes is easier possible than using XML-based schemata for tree-like data structures. For the exchange between Regional Transmission Organizations (RTO), Distribution System Operator (DSO) or Transmission System Operator (TSO), this application and scope of the CIM has proven to be state-of-the-art worldwide and is mandated in some countries (like in the US).

The third main application deals with using more tightened CIM models and structures than using the common data model and pre-defined serialization types. Within the Interface Reference Model (IRM) for the SIDM Systems part of the IEC 61968, pre-defined processes and payloads exist which ca be used as blueprints to start on using standards-compliant processes and XML schemata. Apart from the data model, basic payloads and processes are defined, as seen e.g. in IEC 61968-9 for metering. Through XML namespace mechanisms, additional, utility- or vendor-based semantics can be added. Furthermore, the GID providing a generic interface definition for getting data based on CIM semantics but with no special payload attached. The idea of the GID is more focused on the the status of the data (historical data, alarms and conditions, high-speed data access) and using it alongside CIM semantics. Therefore, the use of mass-data generic interfaces is lees profiled and the other use cases.

The fourth main use case deals with using CIM for market communications and data exchanged, covering European-style and US-style markets with new objects extending the original EMS purpose. The market communication is a regulatory issue with bodies making certain aspects of the standards e.g. certain profiles, mandatory and therefore, a European-profile might just exist for technical data exchange on the European Network of Transmission System Operators for Electricity (ENTSO-E) level but not between TSO and DSO on national level in the member countries.

Having provided an introduction into it's applications, the next section are going to provide information how the CIM is standardized and what standards are closely related to the CIM.

1.6 Standards around Common Information Model

1.6.1 IEC 62357: Seamless Integration Reference Architecture

This standard is one of the more comprehensive ones in the TC 57 area of responsibility. It combines many different IEC standards for electric power system management and associated information exchange into one extensive reference architecture. According to [61], the standard has three objectives:

- Provide a reference architecture to show how the various existing standards within the IEC TC57 relate to each other.
- Identify areas where harmonization between TC 57 standards is needed
- Define a vision for future standardization work

These objectives where the original motivation for creating the SIA, and are mostly for standardization purposes. But a closer look reveals also chances and possibilities outside the standardization process. Before the creation of this standard, no official view on the standards and its connection points as well as the viewpoint from TC57 itself exist. This informations can be useful for integrating standards into the corporate processes and standardization efforts.

The basic structure is illustrated in figure 1.5. The structure of the architecture can be divided into three parts which are represented through the rectangles and marked with A to C. The different layers and pillars hereby show a further structuring mechanism.

The layers included in Part A are mainly for business integration, data definitions and application. In this part, the CIM plays the main role because it is the only standard in this abstract layer. The first layer covers the standards for the integration of external applications, whereas layer 2 and 3 consider the data concepts and interfaces for the specific applications named in layer 4 which should be provided by the CIM. These parts can be implemented in CIM seamlessly. For the transformation of the abstract concepts to the field device communication concepts, layer 5 is responsible.

The standards used in part B of the figure 1.5 are for the direct communication with field devices. These parts are out of the scope of the CIM. But of course the interfaces between layer 5 and 6 have to be harmonized to reach a interoperable seamless integration. Part C holds cross-cutting standards like the important IEC 62351 standard which is further described in chapter 1.6.7.

Within the SIA, several other recommended Smart Grid standards exist. We will provide a short overview on them in them following sections, varying in length with their respective importance regarding the CIM.

1.6.2 IEC 60050 - International Electrotechnical Vocabulary - Electropedia

Electropedia (also known as the International Electrotechnical Vocabulary - IEV) is the world's most comprehensive online electrical and electronic terminology

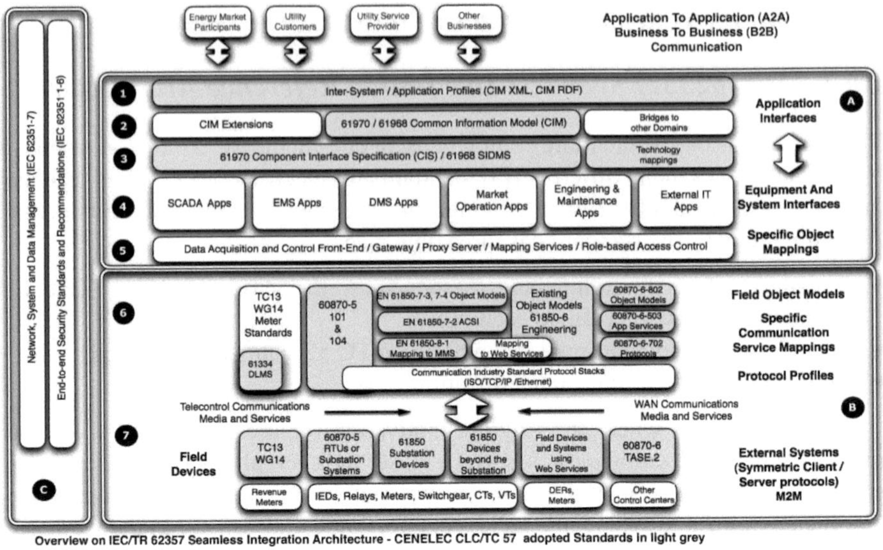

Fig. 1.5 Annotated overview of the IEC TR 62357 Seamless Integration Reference Architecture based on [61]

database containing more than 20.000 terms and definitions in English and French organized by subject area, with equivalent terms in various other languages, whereas coverage varies by subject area like optics, mathematics or generation technology.

Electropedia is both produced and maintained by the IEC and contains all the terms and definitions in the International Electrotechnical Vocabulary or IEV which is published also as a set of publications in the IEC 60050 series that can be ordered separately from the IEC webstore but is always available free of charge at the web site `www.electropedia.org`.

1.6.3 *IEC 61850 - Communication Networks and Systems in Substations*

Originally designed with the designation of enhancing the interoperability between systems from different vendors at substation level, the standard family has outgrown its original purpose and has become an automation and control standard at the field level. Providing a abstract data model, less object-oriented and IT-driven than the CIM but still useful with tree-like structure for navigation, an abstract communication system interface provides the technical mapping from the communication standards which ca be exchanged as time and technology evolves to the core functions, status points and attributes of the physical systems modeled.

1.6.4 IEC 60870 - Telecontrol Equipment and Systems

Focusing on the actual communication link between SCADA and substation, these communication protocol standards facilitate control center data exchange and can be applied to any other domain with the same requirements, e.g. power plants or DER. The standards from this family define the electric power systems status and the control messages to be sent between Control Centers (CCs) of different DSOs and TSOs.

1.6.5 IEC 62541 - OPC Unified Architecture

The OPC Unified Architecture(OPC-UA) is a new automation paradigm and standards series replacing the old Microsoft based and driven OPC Data Access (DA). It provides several layers which can be used in context with the IEC 61850 and CIM and provides two main serializations, a binary and a web service one.

1.6.6 IEC 62361-100 Harmonization of Quality Codes across TC 57 - Part 100: Naming and Design Rules for CIM Profiles to XML Schema Mapping

This standard was previously known as IEC 61968-14 System Interfaces For Distribution Management - XML Naming and Design Rules [63] and was renamed and repositioned recently. As the name states it defines naming and design rules for the generation algorithm of creating XML schema [66]. It is applied within the TC 57 WG 14 group for developing profiles in the standardization profiles but can also be applied by CIM users in practical application scenarios. This subpart uses the existing UN/CEFACT Naming and Design Rules (NDR) rules if applicable and has numerous reference to the document [112].

1.6.7 IEC 62351 - Data and Communication Security

The IEC 62351 Standard addresses the cross-cutting topic of security aspects in the ICT domain regarding IEC standards. The standard defines it main objective itself as follows [39]:

> Undertake the development of standards for security of the communication protocols defined by the IEC TC 57, specifically the IEC 60870-5 series, the IEC 60870-6 series, the IEC 61850 series, the IEC 61970 series, and the IEC 61968 series. Undertake the development of standards and/or technical reports on end-to-end security issues.

In the context of this book the following sub-standards can be of interest:

- IEC 62351-1: Data and communication security - Part 1: Introduction and overview: Provides a introduction of the overall security domain and various aspects and an overview for the upcoming other remaining parts [36].

- IEC 62351-2: Data and Communication Security - Part 2: Glossary of terms: As in nearly all IEC standards, the Part 2 substandard provides a glossary for the specific standard [48].
- IEC 62351-3: Data and communication security - Part 3: Profiles including Transmission Control Protocol (TCP) and the Internet Protocol (IP): This part describes the use of Transport Layer Security (TLS) to secure TCP/IP based connections like in the telecontrol environment [40].
- IEC 62351-7: Communications and Information Management Technologies - Network and System Management in Power System Operations: The IEC 61351-7 mainly defines models specific to power system operations. Mappings to the CIM are planned for separate publication [60].
- IEC 62351-8 Power systems management and associated information exchange - Data and communications security - Part 8: Role-based access control: As the name states the standard deals with the role based access control. The standard has yet to be released and is currently only available as a draft document [73].
- IEC 62351-9 Power systems management and associated information exchange - Data and communications security - Part 9: Cyber security key management for power system equipment: This part addresses the process of assuring a secure and correct handling of key material as well as additional security objects like passwords. It is a technical specification which provides definitions for asymmetric as well as symmetric key management systems [72].

Because the CIM is in most cases technology unspecified, and a majority of security specifications are technology specific, it is hard to recommend security options and solutions. It is often only possible to give general security advices and follow the technology specific security like WSS in the implementation phase.

1.7 IEC Standardization Organization 57

The IEC[4] is the world leading organization for publishing International Standards in the field of electrical, electronic and related technologies. In the following a short introduction into the IEC, based on the information provided at [76], is given.

The IEC was founded in 1906 and is a not-profit, non-governmental organization which provides a platform to companies, industries and governments for meeting, discussing and developing the International Standards they require. Its members are National Committees (NCs) (like for example the German DKE[5] in Germany). The IEC comprises one member NC per country, each of them pays membership fees and in exchange can participate in IEC work. The NCs appoint experts and delegates coming from industry, government bodies, associations and academia to participate in the technical and conformity assessment work of the IEC. Currently, the great majority of experts come from industry.

[4] See http://www.iec.ch

[5] See http://www.vde.com/en/dke/Pages/DKE.aspx for information on the German Commission for Electrical, Electronic & Information Technologies.

Fig. 1.6 shows the organizational overview of the IEC SMB. The SMB manages the IEC standardization activities and creates TCs for different scopes.

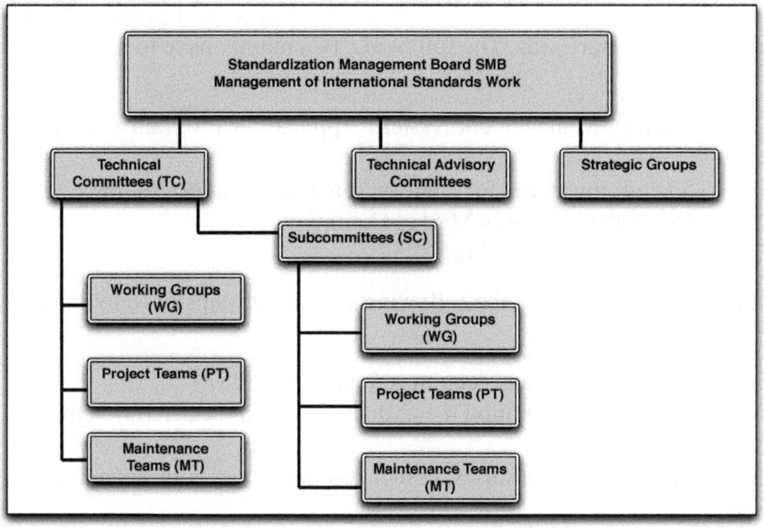

Fig. 1.6 Standardization Management Board Organigram [76]

TCs and SCs are responsible to prepare technical documents on specific subjects within their respective scopes. These documents are then submitted to full member NC for vote with a view to their approval as International Standards. Further on, TCs report to the SMB about their work program and planned liaisons with other organizations. Regarding participating NCs two types of membership exist:

- P-Members (Participating members): must vote at all stages and contribute to meetings
- and O-Members (Observer members): may follow the work as an observer receiving committee documents with the right to submit comments and to attend meetings.

TCs activities need to be approved by the SMB, here the SMB is supported by Technical Advisory Committees and Strategic Groups. On the one hand, Technical Advisory Committees focus on issues such as safety, environment and electromagnetic compatibility. On the other hand, Strategic Groups support the SMB with regard to market needs and potential cross-over of work. SCs might be created by TCs when they have to deal which too many work items. The coordination of the SCs is done by its parental TC. Inside of TCs and SCs the work is done by WGs, Project Teams (PT) and Maintenance Teams (MT).

At the moment, 175 TCs and SCs, 425 WGs about 700 PT / MT carry out the standards work of the IEC. These working groups are composed of people from all around the world who are experts in electrotechnology. In all, some 10 000 experts worldwide participate in the technical work of the IEC.

As mentioned before, the TC 57 itself consists of several groups which have to cope with different standards. The following TCs mainly have to deal with the CIM family standards.

- WG 13 - Energy Management System Application Program Interface (EMS - API)
- WG 14 - System Interfaces for Distribution Management (SIDM)
- WG 16 - Deregulated Energy Market Communications
- WG 19 - Interoperability within TC 57

Apart from that, currently the following working groups have to be regarded as active:

- WG 3 Telecontrol & Teleprotection standards for 60870
- WG 9 Distribution automation using distribution line carrier systems using IEC 61334
- WG 10 IED communications & associated data models in power systems with 61850
- WG 15 Data & Communication Security for IEC 62351
- WG 17 Communication systems for DER using IEC 61850-7-420
- WG 18 Communication systems for Hydroelectric power plants using IEC 61850-7-410
- WG 20 Planning of power line carrier systems using IEC 60495, IEC 60663
- WG 21 System interfaces and communication protocol profiles relevant for systems connected to the Smart Grid (NWIP)

Those committees are currently active and keep maintaining the existing standards as well as providing the new work item proposals. As time of writing this book, the working group 21 for system interfaces and communication protocols relevant for systems connected to the smart grid was still under formation and had no first meeting. It is expected that this group will have a strong background from experts of all the other groups just like the WG 19.

Figure 1.7 shows the work of the TC 57 in the historic context how the EPRI groups influenced the CIM and which WG existed in the very scope of the TC 57 back then.

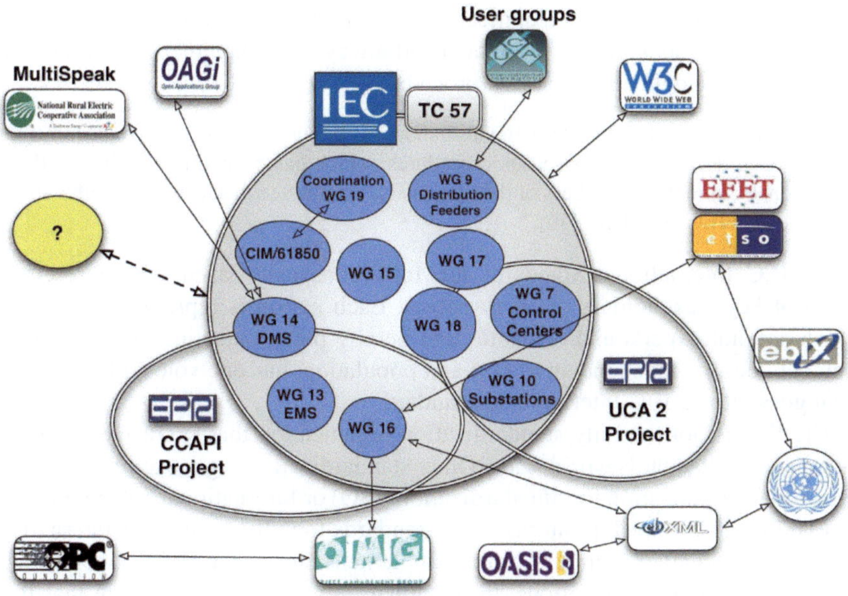

Fig. 1.7 IEC/TC 57 in the historical context of former committees and organizations [12]

1.8 Standards Development and Lifecycle Processes

This section introduces IT standardization as well as standards related development and lifecycle processes. The descriptions follow the general IEC procedures and directives available at [76].

1.8.1 Standardization

Standardization regarding information technology aims at defining and specifying e.g. data models, protocols and techniques to support data exchange between applications as well as integration of applications and devices. The aim is to enable the use of devices and applications of different vendors as well as to support the exchange of data between business partners. The principal outcome of standardization efforts is the development of standards and related documents.

> "According to [75] a standard is a document, established by consensus and approved by a recognized body, that provides, for common and repeated use, rules, guidelines or characteristics for activities or their results, aimed at the achievement of the optimum degree of order in a given context."

The IEC focus on the development of International Standards for the energy industry and, thus, standards development and life cycle management are described in the following as defined by the IEC procedures.

> According to the IEC an International Standard is defined as "a normative document, developed according to consensus procedures, which has been approved by the IEC National Committee members of the responsible committee in accordance with Part 1 of the ISO/IEC Directives [79]."

All IEC International Standards are fully consensus-based and represent the needs of key stakeholders of every nation. Each nation is represented by its NC which appoint experts and delegates and hereby participates in IEC work. Every member country, independent of size and population, has one vote and influence in what goes into an IEC International Standard.

The IEC is not the only standardization organization that exclusively develops International Standards world-wide. Other standardization organizations like the International Organization for Standardization (ISO) or International Telecommunication Union (ITU) develop international standards too. To ensure that International Standards fit together seamlessly and complement each other, IEC cooperates with these global sister organizations. In particular, joint committees ensure that International Standards combine all relevant knowledge of experts working in related areas. Apart from joint committees, at working group level liaisons between other standardization organizations as well as trading organizations and other were established. This aims at avoiding to reinvent the wheel as well as leveraging existing standards and specifications.

Standardization is not only done at international level within IEC. For example, in Europe Standardization is done on National, European and International level. Fig. 1.8 shows the different levels of standardization providing opportunities to different stakeholder for influencing the standards development process.

The European view on standardization as illustrated in Fig. 1.8 comprises three levels. The first level deals with standardization on national level where the corresponding national committee is in charge of the organization and management of national working groups which adopt international standards, develop amendments or new proposal. This national committee usually is member of the IEC and enables national stakeholder to participate in the standardization process. Regarding the CIM, the DKE is the German national committee where its working group K 952 and its subgroups deal with the development of the CIM. On European level, CENELEC is responsible for standardization in the electrotechnical engineering field in Europe. CENELEC mainly adopts international standards according to European needs. On the top level, IEC is responsible for the development of international standards and other publications with international focus. Standardization processes comprises both top-down and bottom-up approaches.

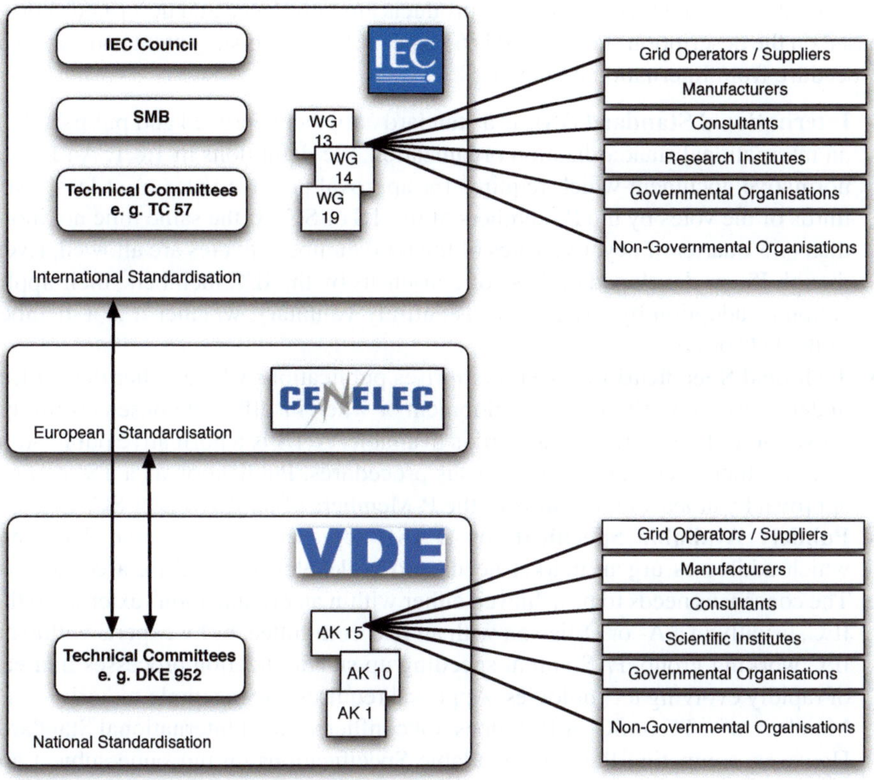

Fig. 1.8 Standardization organizations [12]

1.8.2 IEC Publications

International Standards are the most known documents published by the IEC, but not all standardization efforts lead immediately to the development of an international standard. Apart from International Standards, IEC TCs/SCs develop other types of publications for a specific area of electrotechnology. Publications of the IEC can be assigned to the following two broad categories:

- **Normative publications:** specify agreements on the technical description of the characteristics to be fulfilled by the product, system, service or object in question. For example international standards and technical specifications as well as related documents.
- **Informative publications:** offer background information such as implementation procedures or guidelines like technical reports.

International Standards and other publications of the IEC are the result of full or limited international consensus among the IEC's members (NCs). IEC publications are usually published bilingual in English and French. In the following paragraph, documents developed within the IEC, products are introduced and described shortly based on the descriptions in [79] and [76]. They differ in scope, content and approval procedure from International Standards.

- **International Standard (IS):** is a standard adopted, approved and published by an international standardization organization, see definitions in 1.8.1. An IS is a normative document which requires for approval positive votes of at least two-thirds of the votes by the P-members of the TS or SC. At the same time not more than one-quarter of negative votes of the total number of votes are allowed. Even though IS are developed on base of consensus by the IEC members, their application or adoption by any country is entirely voluntary, whether it is a member of the IEC or not.
- **Technical Specifications (TS):** comprises publications where either the subject under question is still under development or when insufficient consensus for approval of an International Standard is available. A TS is a normative publication and developed according to consensus procedures. Publication as a TS requires approval by at least two - thirds of the P-Members of an IEC TC or SC.
- **Publicly Available Specifications (PAS):** represents a normative document which addresses urgent market needs and is developed based on a consensus. The consensus needs to be achieved either within an organization[6] external to the IEC, usually in a A- or D-liaison with an IEC committee, or by experts within an IEC working group. PAS aim at speeding up standardization processes in areas of rapidly evolving technologies. Approval requires only a simple majority of P-Members of a TC or SC. A PAS does not conflict with an International Standard. However, competing Publicly Available Specifications on the same subject are permitted.
- **Amendments:** are normative documents developed according to consensus procedures which amend, alter or add existing IS. Hereby it changes the technical normative elements of a particular IS. Approval is done through IEC members.
- **Interpretation Sheets:** provides a formal explanation to an urgent request by a user of a standard (e.g. testing laboratory, certification body or manufacturer). Interpretation Sheets aim at quickly providing necessary information to users to facilitate the use of standards. Usually, information provided in Interpretation Sheets are included in the next edition of the related standard.
- **Technical Reports (TR):** are informative publications that comprises background information like data obtained from a survey carried out among national committees, data of work in other international organizations or data on "the state of the art" in relation to standards of national committees on a particular subject. TRs shall not contain matter implying that they are normative. A TR is approved by simple majority of P-Members of an IEC TC or SC.
- **Technical Corrigenda:** are normative publications that are developed to either correct or update information within IEC documents. In the first case, a Technical Corrigenda is published to correct a technical error or ambiguity within an IS, a

[6] For example manufacturers or commercial associations, industrial consortia, user group and professional and scientific societies.

TS, a PAS or a TR to avoid incorrect or unsafe application of the publication. In the second case, information that has become outdated since publication of the standard is updated, as long as the modification has no effect on the technical normative elements (see definition in [80]).

- **Guides:** are usually informative publications providing rules, orientation, advice or recommendations relating to IS and conformity assessment. Some guides are normative, see [76]. Approval of a guide requires at least 75% positive votes of the member bodies casting a vote.
- **Technology Trend Assessment (TTA):** focus on highlighting certain aspects of a technology that might become an area for standardization in the near-to-medium term. They aim at supporting global collaboration on standardization questions during the early stages of technical innovation. TTAs are developed usually within pre-standardization work or research and provide information regarding the state of the art or trend in emerging fields.

1.8.3 IEC International Standards Development

The development of standards is (in general) a time consuming process from the first draft document to the approved standard. Within the IEC the development of an international standard is regarded as the result of an agreement between the participating National Committees. According to [79] the development of a new IEC international standard comprises seven stages, see Fig. 1.9.

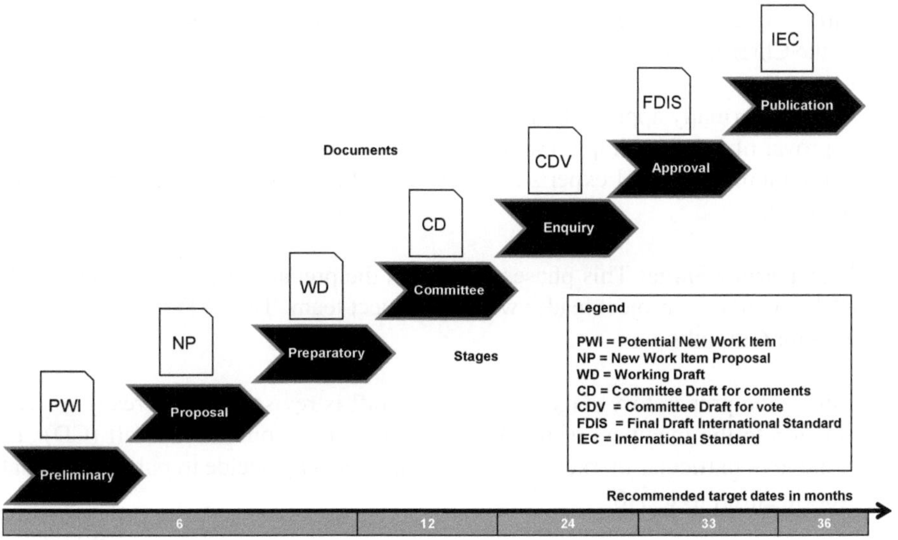

Fig. 1.9 International Standard project stages based on [79]

Fig. 1.9 shows the seven stages and corresponding documents as well as the recommended time frame for developing an International Standard. From the *Preliminary Stage* to the *Publication Stage* several documents (from the Potential New Work Item to the final IEC international standard) are developed, continuously reviewed/discussed/voted and finally agreed on/approved by experts of the National Committees. As illustrated in Fig. 1.9, a standard development project should last about 36 months from its beginning till its final publication as International Standard. Within this process consensus is regarded as an essential principle and necessary condition for the preparation of an International Standard. Here consensus requires the resolution of substantial objections.

In the following the seven stages are described according to [76]:

- **Preliminary Stage:** This is the first stage of the process where usually a Potential New Work Item Proposal (PWI) and the development of an initial draft is done. Activities at this stage comprise preliminary work such as better definition of a project for new work, data collection or round-robin tests. In addition, this stage is applied for work items where no target dates can be established.
- **Proposal Stage:** The proposal stage comprises the development of a New Work Item Proposal (NP) either for a new standard, a new part of an existing standard or a TS. Instead of NP also the abbreviation NWIP is used.

  Such a proposal usually originates from industry via a NC, which informs the members of the appropriate TC or SC using a particular form.The development of a NP may be done by a national body, the secretariat of that technical committee or subcommittee, another technical committee or subcommittee, an organization in liaison, the technical management board (or one of its advisory groups) or the Chief Executive Officer.

  A NP is normally approved after a 3 months commenting and voting period. For approval of a NP a simple majority of the committees' P-members as well as a minimum number of experts nominated by P-members supporting the NP are required.[7]

- **Preparatory Stage:** This phases comprises the preparation of a Working Draft (WD) often by a project leader within a project team. This WD is usually available for 6 months.

  During the preparatory stage the working draft is revised till it is ready for circulation to the members of the TC or SC as a first Committee Draft (CD). To respond to particular market needs the committee may decide to publish the final working draft as a PAS.

[7] Committees with 16 or less P-members need a minimum of 4 experts and committees with 17 or more P-members a minimum of 5 experts.

- **Committee Stage:** At this phase, the document is submitted to the NCs as a CD for comment. The CD is usually available for 12 months.

  At the committee stage, comments from national bodies are taken into consideration, aiming at reaching consensus on the technical content. National bodies are requested to carefully study the CDs and submit all relevant comments at this stage. As soon as it is available, a CD is circulated to all P-members and O-members of the TC or SC for consideration, with a clear indication of the latest date for submission of replies. Here, a period of two to four months is available for national bodies to comment.

- **Enquiry Stage:** The enquiry stage is the last stage at which technical comments can be taken into consideration. Here, based on the comments for the CD of the preceding stage, a bilingual Committee Draft for Vote (CDV) is developed and submitted to all NCs for a five-month voting period.

  The CDV is approved if a majority of two thirds of the votes cast by P-members is in favor and if the number of negative votes cast by all NCs does not exceed one quarter of all the votes cast. A revised version is then sent by the secretary to the Central Office within four months for Final Draft International Standard (FDIS) processing. In case a CDV is approved with no negative votes and there are no technical changes, it can proceed directly to publication. Also, if the document should become a TS (and not an IS), only a majority of two thirds of the votes cast by P-members is needed.

- **Approval Stage:** Within the approval stage the FDIS is circulated to the NCs for a two-month voting period. At this point each NC's vote requires to be explicit, only positive, negative or abstention is allowed. In addition, no comments are allowed with a positive vote.
  Approval of an FDIS requires that a majority of two thirds of the votes cast by P-members is in favor and less than 25% of all votes submitted are negative. In case the document is approved, it progresses to the final publication stage. Otherwise it is referred back to the TC or SC to be reconsidered.

- **Publication Stage:** The publication stage is the final stage within the development of an IS at IEC. This stage is entirely in the responsibility of the Central Office and leads to publication of the IS. Once a final draft IS has been approved, only minor editorial changes are introduced into the final text. Publication of the IS is done normally within 1.5 months of approval of the FDIS.

The development of IS and other documents at IEC are driven by the key principle of international consensus among the IEC's members (NCs). Any member of the IEC may participate in the preparatory work of an IS, and any international, governmental and non-governmental organization liaising with the IEC also participates in this preparation.

In addition a truly IS can be submitted to public enquiry in any of the IEC NCs' countries. Thus, through the democratic tools of consensus and public enquiry, any interested party may speak up and have their say in the development and publication of an IEC IS.

1.8.4 IEC Publications Lifecycle Management

IEC publications undergo several steps from their initial creation over publication and maintenance till their finally retirement. Fig. Figure 1.10 illustrates the basic steps involved within the lifecycle management, hereby it adheres at the procedures common for IEC international standards.

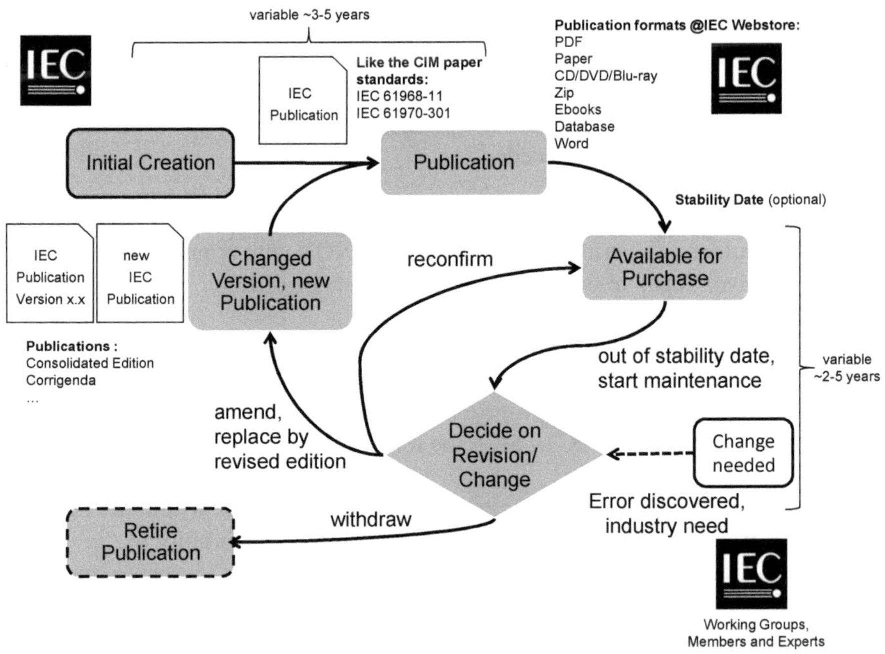

Fig. 1.10 IEC Publications Lifecycle

In the following the basic steps are described:

- **Initial creation:** This is the starting point of any publication. It comprises the several consensus based procedures within IEC, see section 1.8.3, and requires usually three to five years till the initial document is ready for publication.
- **Publication:** A document can be published in various formats, like paper, pdf or ebook. Further artifacts like electronic models might be added and provided. Publications of the IEC can be obtained through the IEC webstore[8].
- **Available for purchase:** After publication documents might get a stability date which defines the period of time for which they can be purchased. The stability date varies from two till five years depending also on the type of publication.
- **Decide on revision or change:** When a document passes its stability date then replacement of the document according to its maintenance cycle is due. At this moment in time, a decision has to be taken either to reconfirm, amend, publish a new version or retire the publication. Apart from the planned maintenance cycle, a decision on revision or change of a document can always be triggered if an error is identified or because of changed business requirements. Errors or industry needs are communicated usually via the NCs to the corresponding IEC TCs and SCs.
- **Changed version, new publication:** Depending on the required amendments a changed version of an existing publication or a new publication might be developed. Within this several types of documents might be developed, see section 1.8.2. As a result updated documents or even new documents will be made available for purchase at the IEC webstore and a new maintenance cycle will start.
- **Retire publication:** This steps comprises the end of the lifecycle of a document. After withdrawal a document won't be available for purchase anymore.

After development and publication of an IEC document its adoption, regardless of the type of document like TS or IS, remains voluntary by any country, whether it is a member of the Commission or not.

Participation within standardization is possible during the different steps, especially at creation, decision on revision and development of new or amendment of existing versions. Companies willing to influence the development of IEC publications have different possibilities to participate. On the one hand, direct participation by joining IEC working groups through nomination by their NC or participation at national level by joining national IEC mirror committees is possible. On the other hand, indirect participation through associations, user groups or other standardization bodies and working groups with established liaisons with the IEC can be done. Figure 1.11 shows the different levels of standardization from a German perspective, hereby international (top-level), European (middle-level) and national (low-level) are differentiated. From top to bottom the level of detail and amount of documents increase due to operational requirements, like specific regulation guidelines, that have to be considered.

[8] See `http://webstore.iec.ch/`

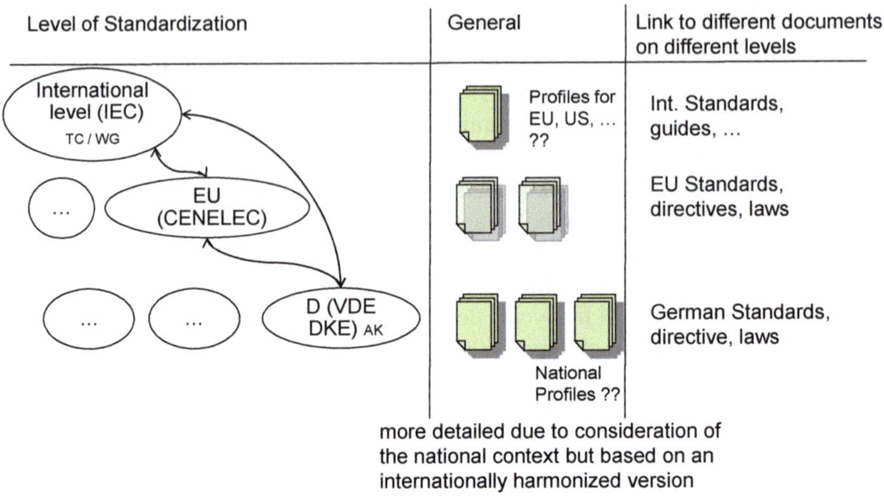

Fig. 1.11 Levels of standardization

1.8.5 CIM Data Model Development Process

As stated in the sections before several standards and related documents exist. The aim of the IEC 61970/61968 standards families is to support integration of applications and devices and reduce the "distance to integrate". When integrating applications from several vendors, several approaches exist.

If no standards exist custom integration is needed which in general is the most costly one regarding maintenance. Standard based integration promises higher quality and lower maintenance costs in the future but requires the availability of existing standards. Within standards based integration mainly three types of standards can be distinguished:

1. standards aiming at syntactic interoperability, providing interface descriptions
2. standards focusing on semantic interoperability, in addition to syntactic interoperability a common semantic data model is provided
3. plug and play standards, (ideally) no additional customization is needed applications just have to provide fully specified interfaces and exchange of data will be able

The CIM focus on semantic interoperability and apart from providing a common semantic data model within IEC 61970, 61968 and 62325 a set of profiles, interface descriptions and use cases are provided. The development of the IEC CIM data model comprises several steps from its initial creation till the provisioning of a new version of the electronic model.

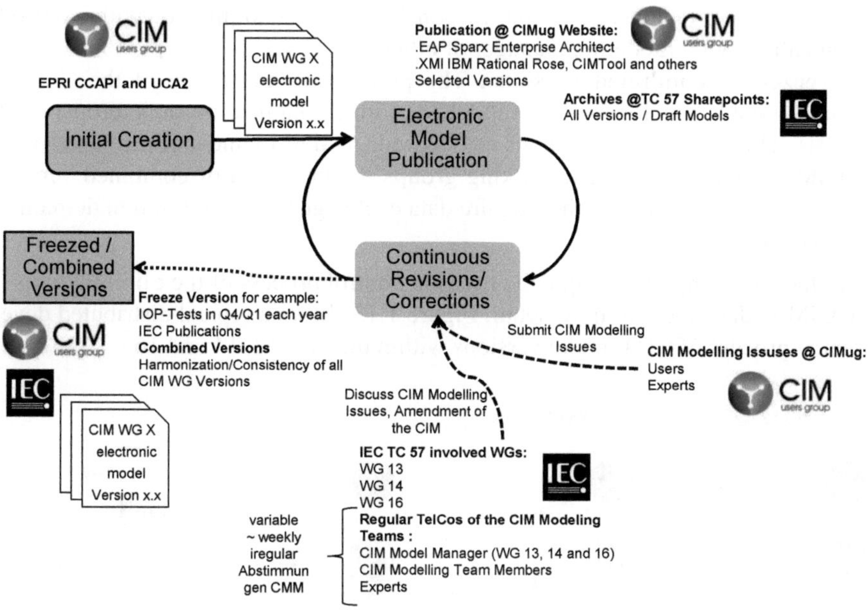

Fig. 1.12 CIM electronic model lifecycle

In the following the basic steps within CIM model development are described:

- **Initial creation:** The first version of the CIM was developed based on the contributions of two EPRI projects, CCAPI and UCA2, see section 1.4. Based on this several version were developed till now.
- **Electronic model publication:** Documents and artifacts regarding the CIM are provided on two different sides. On the one hand, the current as well as selected previous versions of the electronic model of the CIM are hosted at the CIMug website[9]. Here it is provided as Enterprise Architect model, whereas several other formats like IBM Rational Rose are also available. On the other hand, the IEC TC 57 hosts all standards related documents as well as further draft model versions.
- **Continuous revisions or corrections:** The CIM is continuously evolving through revisions and corrections. Input and requirements regarding changes of the CIM are provided either through the CIMug or corresponding IEC working groups. Members of the CIMug can submit modeling issues through the CIMug website which will be discussed at IEC modeling team meetings and teleconferences. Modeling issues are discussed and further activities are performed within modeling team conference calls on a regular basis, usually weekly. The modeling team is currently comprised of IEC experts of the TC 57 working groups 13, 14 and 16. Within CIM data model development, WGs are responsible for certain parts/packages of the CIM whereas certain parts are used by several WGs. Each WG

[9] See `http://cimug.ucaiug.org`

has its modeling manager who is responsible for the coordination within the WG and aligns with modeling manager of other WGs.

- **Freezed or combined versions:** The publication and usage of the CIM requires stable versions of the model for further publication or interoperability tests. Therefore model versions are freezed. As development of packages of the CIM is done by different working groups the provision of combined versions for use within scenarios that require data exchange between different domains is necessary.

In the following praragraph, the revision control process of the different parts of the CIM is described in more detail. Figure 1.13 shows a typical distributed development process of CIM model versions within the WG 13, 14 and 16.

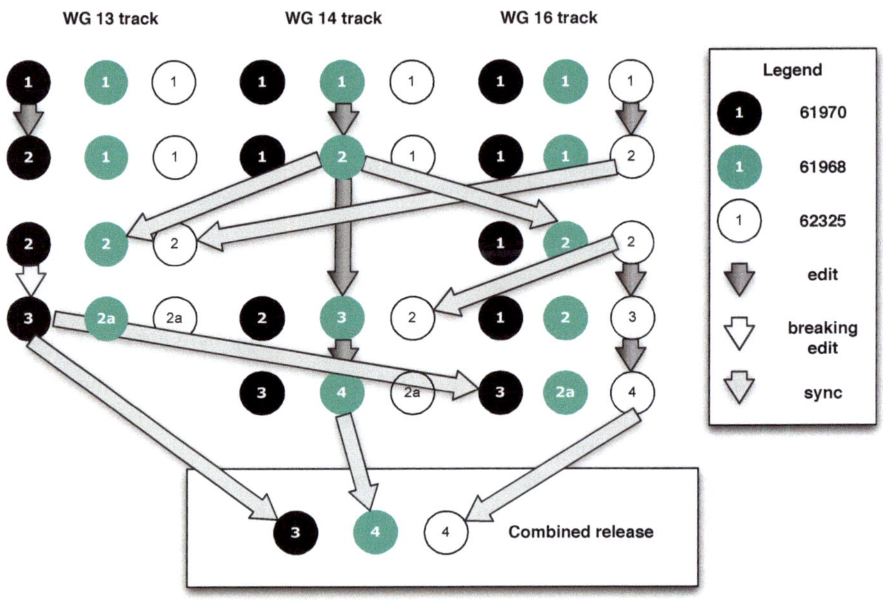

Fig. 1.13 Distributed development process of CIM model versions [9]

Each WG works on their corresponding part of the CIM (see three lanes) and manages its own release number as well as decides on enhancements and amendments (edit). From time to time a synchronization with other WGs might be necessary as changes might affect several WGs (synch). Figure 1.14 shows possible synchronizations between the different WGs.

Due to dependencies between packages and classes as well as different priorities within the WGs no continuous synchronization between all parts of the CIM is done. By changing and editing a part of the CIM within a WG the different model versions might become out of sync (breaking edit, Fig. 1.13). Thus, combined releases are necessary to provide a combined version of the whole CIM which is in sync (combined release).

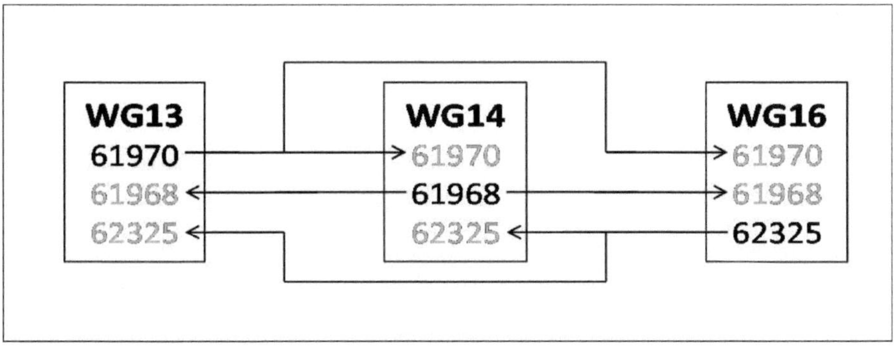

Fig. 1.14 Possible synchronizations between WGs [9]

A combined release version is produced by concatenation of the corresponding CIM model version of the WGs 13, 14 and 16 with regard to the according IEC Standard IEC 61970, 61968 and 62325. Each WG manages its own model version for a selected part of the CIM which corresponds to the IEC Standard in charge. This model version is build based on model part, major release and minor version, see Fig. 1.15.

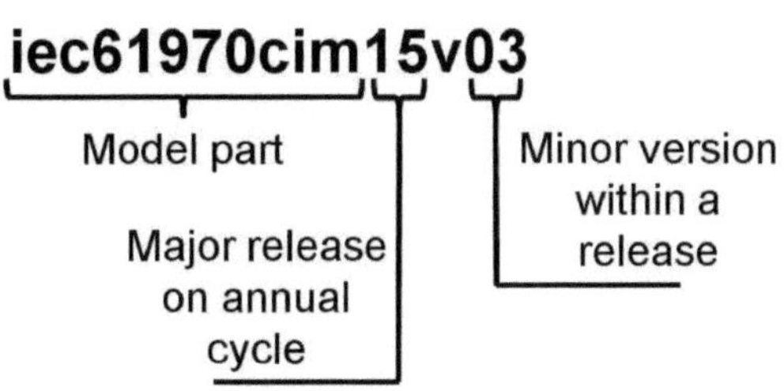

Fig. 1.15 Example for IEC 61970 model version [98]

For example, given the following model versions iec61970cim15v03 (WG 13), iec61968cim11v02 (WG 14) and iec62325cim01v00 (WG 16). The combined model version would be iec61970cim15v03_iec61968cim11v02_iec62325cim01v00.

Chapter 2
Basic Technologies

The CIM is described in the Unified Modeling Language (UML) which is specified by the Object Management Group (OMG). To understand the CIM it is required to have basic knowledge of UML. Furthermore the application of the CIM usually involves other description languages like the eXtensible Markup Language (XML) or the Resource Description Framework (RDF) and it is realized within Service-oriented Architectures (SOA) with corresponding platform support by so-called middle ware. This chapter describes the basic prerequisites to understand and apply the CIM. At first the necessary UML basics are being facilitated, secondly XML as a method of serialization and instantiation is outlined and RDF is briefly introduced in the third section of this chapter. Finally, Service-oriented Architectures as an environment for the integration of applications with the CIM are presented.

2.1 UML Basics

The development of large systems is a complex issue that is executed within a number of different activities. An approach which is often used to cope with the complexity is to model the system under design. A model which is created in a modeling process describes selected elements of the real world with their attributes which are relevant for further considerations. It depicts real world entities and does not describe them in full detail – just to the degree necessary to describe the aspect which is to be examined. These models can be used in the same manner as architectural blueprints in the domain of house constructions. Using abstraction helps for example to identify commonalities, problems and ways to improve what is built or shall be built. The CIM in particular considers various entities in the energy domain and captures them in terms of a data model.

Models can be described in different notations and representations, but in the context of information technology the Object Management Group (OMG) Unified Modeling Language™(OMG UML®) is the most widespread notation in the context of software-intensive systems and applied in various domains. It aims to serve as a means to the description of software-intensive systems. With its scope and its

M. Uslar et al.: The Common Information Model CIM, POWSYS, pp. 47–74.
springerlink.com

developments up to now, it is in use by several groups of people, like for instance software architects, software developers or business analysts. Among these different groups of people, UML serves as a common language and facilitates the understanding of the model. Moreover, a unified and interoperable model allows to use advanced tools that realize the benefits of syntactic as well as semantic specification – for example in the form of implementation assistance.

The UML defines different types of diagrams to graphically model systems which serve as views on the model. This allows to express different aspects that are relevant to system development, as for example static structures, dynamic behavior and interactions. Though UML defines graphical representations of elements it does not specify specific steps how to model, what to model or how to implement what has been modeled. This means it is clearly up to the user of the model to interpret the information according to the domain and application area.

2.1.1 The UML Specification

A clear reference for UML is of course the specification itself. UML evolved since the mid 1990s and by now (2011) the latest stable version is 2.3. It is being developed and standardized by the Object Management Group (OMG) as well as in ISO/IEC 19501 (in version 1.4). The UML specification describes how the language can be applied, in terms of how the language features can be combined to form valid descriptions. It is separated into four parts which are:

UML Infrastructure: The UML itself is described using specific elements, which are described within the UML Infrastructure. This approach to create the model that describes UML is called meta-modeling. The infrastructure specification defines a closed set of elements on which other models as the UML Superstructure can be built. This includes for example primitive types, abstractions and constructs.

UML Superstructure: A formal description of how the UML is built up is given by the UML Superstructure. This description is done with the concepts of the UML Infrastructure. The superstructure document is the most comprehensive document from this series and probably the most important to consult for UML users.

Object Constraint Language: The Object Constraint Language (OCL) is a formal language which allows to express constraints for model elements. It can be used to restrict the possibilities UML offers, as for example to constrain values for attributes or objects.

Diagram Interchange: Last but not least, the diagram interchange specification defines how UML models can be shared between different modeling tools.

Besides the elements already defined in the superstructure, there are also mechanisms to extend the language. In order to work with the CIM it is only necessary to understand a subset of the UML language features which are described in the UML Superstructure specification. These are mainly so-called class diagrams and package diagrams that are used to depict the structure of information. Besides these two diagram types, there are some more diagram types to prepare information around systems analysis and design. As these two diagram types are the most important ones in the context of this book, a basic explanation of them as well as basic UML concepts will be subject of the following section.

2.1.2 UML Class Diagrams

Class diagrams are one of the simplest and common diagram types defined in UML. They describe classes and their relations, which is quite close to most object oriented programming languages and human perception of things, and so it is comparatively easy to grasp them intuitively. The versatility of class diagrams allows them to be used for different different purposes. However, there are a few things to know about the elements used in class diagrams. As the name implies the most important elements are classes which are considered in more detail firstly.

2.1.2.1 Classes

A *class* is an abstraction that describes, based on [96], "a set of *objects* that share the same specifications of common features, constraints and semantics". That means objects are instances of classes and their corresponding class specifies their structure and behavior.

Classes are represented by rectangles which are divided by horizontal lines into compartments. Each of the, by default three, compartments, holds specific information of the class. The first one is designated for the name, the second one holds the attributes of the class and the third on its operations (see sections "Attributes" and "Operations"). Compartments can be hidden, e.g. if they are not relevant to be considered for a certain aspect, to make diagrams more readable. Looked at another way, this means it can not be concluded that these hidden information are not specified if they are just not shown. Often, especially in early conceptual development phases, classes are only specified by their name. They are typically labeled with a noun in its singular form and start with a capital letter. An example of a simple class can be seen in figure 2.1.

Fig. 2.1 Example of a UML class named "WindEnergyPlant" and an Instance (Object)

Object

Objects are created based on the description of classes as instances of these. Each object has its own identity and is depicted with a similar shape. Objects can have their own names (but must not) which are followed by a colon and the class name to identify them. To distinguish objects from classes more easily, their name is also underlined.

According to the previous example (see figur 2.1 a specific instance of `WindEnergyPlant` is identified by `WEPBorkum1 : WindEnergyPlant` or it could also be identified by `:WindEnergyPlant` in case it is an anonymous object.

Abstract classes

The paragraph before did not state that all classes can be instantiated but just that every object is an instance of a class. In general, most classes can be instantiated but there is one exception: the abstract class. These classes are typically used to logically group functionalities and are often consciously specified incomplete. In order to instantiate them, they must be extended by another class, which will be described later in section 2.1.3. Abstract classes can be recognized by the italicized font in which their name is written.

Attributes

Attributes define the characteristics concerning the state of classes in more detail. They are defined with at least a name and can have a type, separated by a colon, that describes/restricts allowed values for instances of this class. This could be primitive types which are usually built-in types in an implementation language as for example integer values or strings as well as complex types. A class itself is such a complex type and can be be used as a type for attributes. In this connection it is also possible to specify enumerations which are specially marked classes (see section 2.1.5) that allow to specify a limited set of data values (so-called enumeration literals).

Regarding the attribute name, it is common, and also suggested in the UML specification [95], to begin with a lowercase letter. Attributes can also have a default value which is indicated by an equal sign and the assignment of a value in quotes after the attribute type.

Figure 2.2 gives a detailed view on the class `WindEnergyPlant` with its attributes `accessDoor`, `power`, `manufacturer` and `rotors`.

In this example, an instance of a wind energy plant can have one and up to two access doors and an integer value that represents the power of the unit. Moreover, it has a string that represents its manufacturer (by default "Unknown") and it can have from two up to infinite rotors.

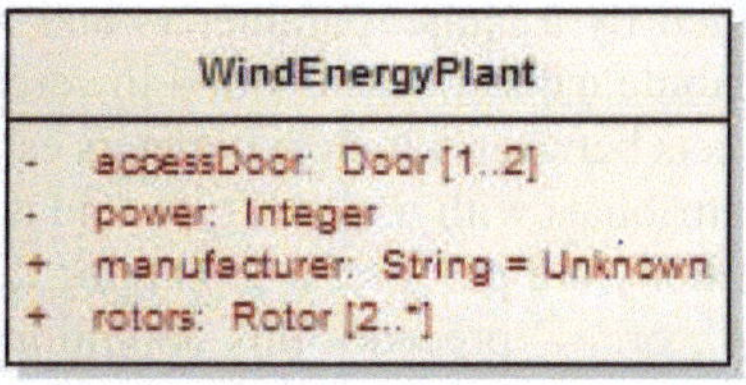

Fig. 2.2 UML class WindEnergyPlant with attributes

Visibility

As also depicted in figure 2.2, preceding characters can be used to indicate the attribute's visibility. In this context the characters '+', '-', '#', and '∼' are allowed shortcuts to represent *public*, *private*, *protected*, or *package* visibility respectively. Throughout the CIM mainly public attributes, which are accessible from other classes, are used.

Multiplicity

Beyond that, attributes have multiplicities that were already included in the previous definition. Multiplicities define how many instances of an attribute type can or are being created in respective objects.

The multiplicity is defined by a lower bound value and an upper bound value that describes how many instances must exist and how many instances may exist respectively. This means that possible values range from 0 to infinite (depicted by a '*'). The values for the bounds are specified between square brackets, beginning with the lower bound value, followed by '..' and the upper bound value. This means for the example in figure 2.2, that a wind energy plant must have at least 2 rotors, but it can for this example, also have up to infinite rotors.

To ease the definition of attributes and multiplicities there are some conventions. Hence a fixed multiplicity of, let us say 5 – meaning [5..5] – can be specified by [5]. In case no multiplicity is specified, a multiplicity of [1] is assumed. If the infinite multiplicity '*' is used solely, this implies a lower bound value of 0.

In addition to the preceding text, there are more options to dwell on attribute definition. Because of the focus of this book, please refer to the specification [95, 96] or technical literature like [99] for more information on that, if necessary.

Operations

Besides the state of classes which is described by attributes, *operations* describe how behavior of classes can be invoked. Operations are not in the focus of the CIM, but as they are basic elements of classes, they will described in brief as it may help to understand the basic concepts.

An operation is specified by a name, parameters and a return type. Operation names should, just as attribute names, begin with a lowercase letter. The operation name is followed by brackets between which parameters can be specified. These can be defined analogous to attributes with a name, type and default value. Finally, the return type and its multiplicity are put after a separating colon. As with attributes, a character of '+', '-', '#', or '∼' precedes this definition to represent the operation's visibility. An example of how a class with operations looks like is shown in figure 2.3. In this figure a private operation `getWindspeed` is defined that returns an integer value and has no parameters. The operation is defined to be private so that it can only be accessed internally from other operations in this class. Another operation `turnRotor` is defined in this class that takes one integer parameter (which should be the angle) – this operation does not return anything.

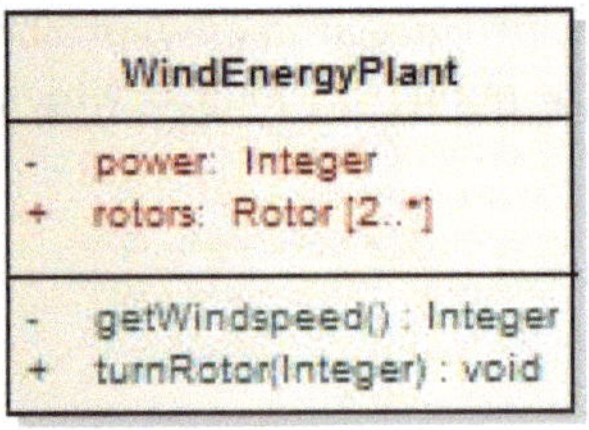

Fig. 2.3 UML Class with Operations

Operations can also be abstract and so they can only be implemented by a subclass. This is graphically depicted by their description written in italics. After having examined the elements of single classes in detail, the relations between multiple classes will be part of the following section.

2.1.3 Relations between Classes

Modeling only classes limits the description of a system to single parts without connections. To take this into account UML provides different kinds of relationships between classes. These are *Dependency*, *Association*, *Aggregation*, *Composition* and *Generalization*. UML is open for interpretation by both, the authors of the model as well as by its readers. Especially in the context of relationships it is often not clear which kind of relation is the most appropriate. So it is important to make sure that readers and model authors have the same understanding of what is meant with the kind(s) of relationships used.

Association

Associations specify "a semantic relationship that can occur between typed instances" [96]. So an association has at least two ends and is defined between two classes which are source and target for an association and equal partners in this association. The source and the target of an association can also be the same. The intention of associations is to explicate longer lasting relationships.

Associations are graphically represented by solid lines between the involved classes. To indicate the navigability of associations they can have arrows at each end that represents the direction of the association. In practice it is unusual to specify arrows on both sides to express a bidirectional navigability as it is mostly the case. With that in mind, it is also not possible to differentiate between an unspecified navigability (no arrows) and a bidirectional navigability (no arrows). For this reason it is possible to explicitly express a non navigable direction by an 'X' on the associations line at the corresponding end.

The associations can also have names, usually including verbs, that describe the relations between the participating types in more detail. This information is typically only used to ease the understanding of the model and does not have any effects on the implementation. Additionally, the name can have a small solid triangle attached to show how to read the association name with respect to the associated types. This indicator does not have any influence on the navigability as could be suggested but is just to assist the reader. Each end of an association can moreover have a role name attached that describes the role of the participating classes in their relation. They are properties of the relation or its participating classes and they can, as with attributes, also be constrained in their visibility by using one of '+', '-', '#' or '~'. Roles are properties of classes and are usually realized as attributes of the respective class (there are exceptions which this chapter cannot explain in detail, see [96] for details). Thus, it is an alternative way to specify an attribute with a complex type – it depends on the author of the model to choose the expression which is useful.

To limit the number of instances of this relationship, multiplicities can be defined at each end of the relation. The definition of multiplicities follows the same scheme as with attributes (see section 2.1.2.1) with the exception that they are not bracketed. By default an unspecified multiplicity means a multiplicity of 1.

Figure 2.4 shows most of these properties: The transformer class has a directed association to the wind energy plant class with the name "transforms power". A wind energy plant class has an association to a generator with a publicly visible role

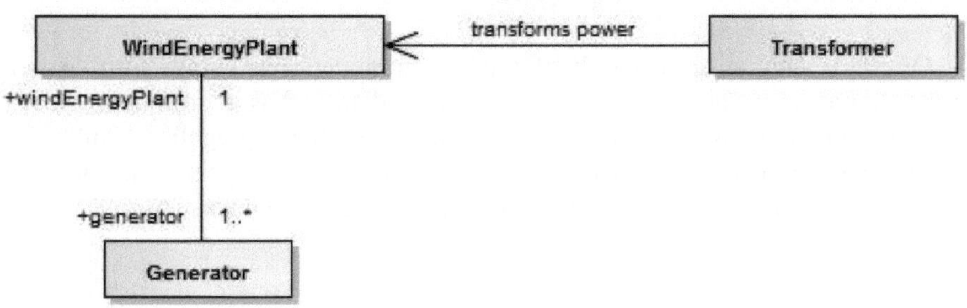

Fig. 2.4 Associations in UML

"generator" and it can have one to infinite generators. A generator on the other hand has exactly one associated wind energy plant available by the publicly visible role *windEnergyPlant*.

Associations can have additional properties, be used with other model elements, be established between more than two classes, etc. Due to the focus of this book, these points are not addressed any further but can for example be found in [96] or [99]. Another kind of relationship is the aggregation which is described in the following section.

Aggregation

A stronger kind of association is the *aggregation* that is a binary association. Aggregation and composition (see next section) are perhaps the two most discussed kinds of association and the usage can vary depending on the modeler.

An aggregation represents a shared association and highlights one end of the association that is strongly using or relying the other end. It is often described as a part-of relation that exists between the involved classes. One major point in this relation that must be stressed is, that both parts can exist without the other one. Insofar such a relationship can be described as a kind of possession, meaning on side owns the other side.

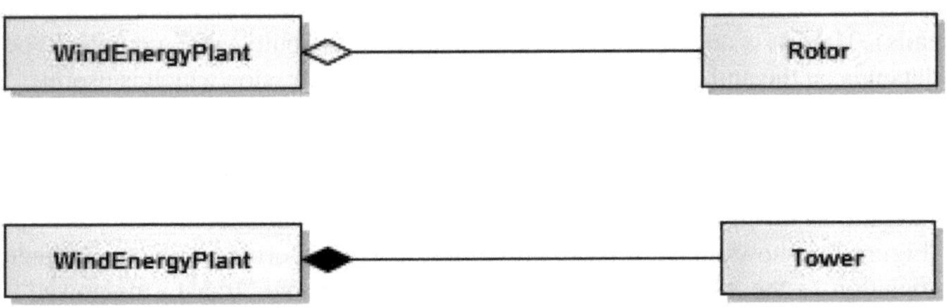

Fig. 2.5 Different Kinds of Associations: Aggregation and Composition

Of course, aggregations can be defined in more detail as known from associations. Aggregations are represented by a hollow diamond as a terminator which is on top of the solid association lines. The side where the aggregation symbol is at defines the owning class – as can be seen in figure 2.5 in the upper part. The aggregation example in this figure shows that a wind energy plant owns a rotor.

Composition

Even stronger kinds of associations between two classes can be represented by composite associations – so-called compositions. A composition is another kind of aggregation that also highlights one side of the relation, and can be explained as a whole/part relationship. Unlike aggregations, the "parts" of the composite has a strong life-cycle dependency between the instances that are more or less "contained" in this kind of association. If the composite is deleted, all of its parts are normally deleted with it.

Furthermore, this implies that "parts" participating in a composition can only be included in at most one single composition association at a time. Compositions can be defined in more detail using the known concepts from associations.

The graphic representation of a composition is a solid diamond crossing the association line on the end of the defining class. An example can be found in figure 2.5 in the lower part. A tower is part of a wind energy plant that could not exist without it.

Dependency

A dependency is a relationship between at least one source elements and at least one target element. It represents a weaker relationship than associations do and signifies "that a single or a set of model elements requires other model elements for their specification or implementation" [96]. This means for instance that the dependent classes briefly interact with the used classes. A dependency can have different manifestations, for example in terms of usage, realization or abstraction.

Concerning the realization, there will generally only be transient relationships established , i.e. the relations do not persist over any longer period of time. Dependencies are shown as a dashed line between the classes and ends with an arrow on the side of the used class. Figure 2.6 shows an example that describes that a wind energy plant depends on maintenance operations. UML brings additional features to specify the kind of dependency in more detail. The mechanism used is called stereotype (see section 2.1.5) and brings with it a number of predefined dependency types, as for example «call», «realize», «instantiate» or «use».

The concept of dependencies can, as well as associations, also be used with elements other than classes – further details can be found in [95] or [96].

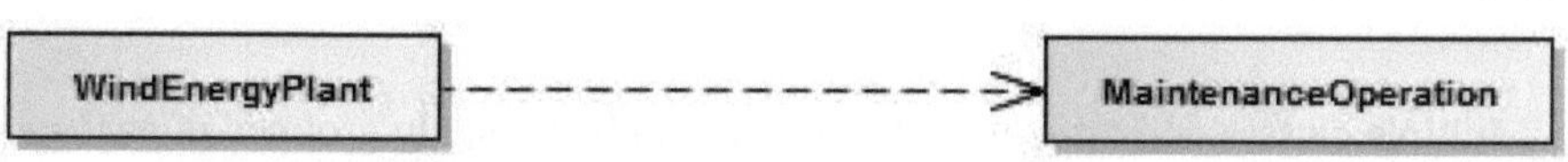

Fig. 2.6 UML Dependency Relationship

Generalization

While preceding relationship types were quite similar, with the *generalization* another aspect comes into play. A generalization is a relationship between a generic and a specialized version of it. It can be best explained with the words 'is-a'. This type of relationship can be used to hierarchically structure classes in order to extract commonalities in more generic classes and specialize them in derived classes. The concept that is illustrated with generalization is called *inheritance* in object-oriented programming languages.

This means that every class that is derived from a class (i.e. a specialization of a generic class) inherits all its features, which especially comprise attributes, operations and relationships, and can further extend the class with its own features or redefine inherited ones to become more specific.

Other classes can then again inherit from these specialized classes. It is also possible that one class can have more than one generalization association and inherits all features from these classes. This concept is called multiple inheritance but is not natively supported in all modern programming languages. The generalization relationship is also the only possible opportunity to subclass *abstract classes* in order to make them instantiable. Generalizations are depicted by a solid line from one class to the other that ends with a hollow triangle on the generic class' side (see figure 2.7). Generalizations are usually drawn from bottom to top for the reason of readability and do not have names or multiplicities assigned. In figure 2.7 a distributed energy resource is defined as a generalized version of a wind energy plant and a hydro plant. Or to say it in other words, a wind energy plant and a hydro plant are specializations of distributed energy resource.

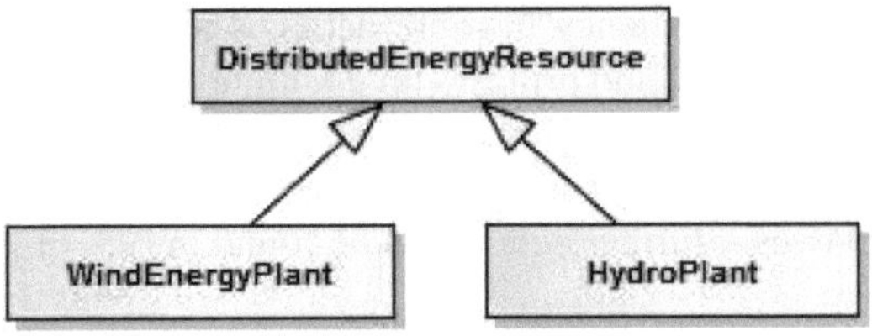

Fig. 2.7 Generalization Relationship in UML to Realize Inheritance

Multiple classes and other model elements can be logically grouped in packages to keep the overall model more manageable and easier to access. Another diagram type – called package diagram – is then used to overview the structure of the overall model. It is briefly described in the following section.

2.1.4 Package Diagrams

With an increasing number of model elements like classes, types and others, the model becomes more and more unclear. To cope with that structuring elements,

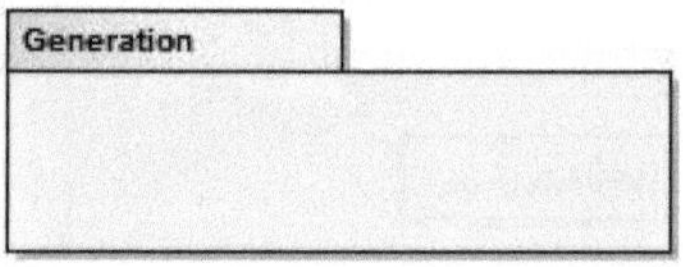

Fig. 2.8 Simple Package "Generation"

called packages, have been introduced. Packages have a name by which the can be referenced and may contain nearly all UML elements like classes are. Moreover, they can also contain other packages to create a finer grained model structure.

In addition to the grouping of related elements, this provides a way to define namespaces for the elements contained. Referencing elements from other packages can be done by their qualified name which results from the package name followed by '::' and the element name. A class `HydroTurbine` in the package `Generation` would have the qualified name `Generation::HydroTurbine` for example. The contents of the package (namespace) can be referred by elements in the same package without specifying the qualified name.

Furthermore, a package, in terms of its function as a namespace, can also import individual or all elements of other packages. The visual representation of packages is a large rectangle with a smaller one attached on its top on the left-hand side. The package's contents must not be shown at all or can only be shown partly. This implies that no assumptions regarding the complete contents of a package can be made by the visual representation in a package diagram – no visible elements do not necessarily mean no contents. A simple package is shown in figure 2.8.

To restrict the visibility of elements, the may be declared as private ('-') or public ('+') as a prefix of their name and so just be used from elements inside the package or also by other packages, respectively.

There are two ways to show elements within a package in a package diagram. The elements can be shown as rectangles inside the large rectangle or they can be drawn outside the package and be attached to it. This is realized with solid lines from the elements (outside of the package) that have a circle with a plus sign at the end close to the package to denote the containment. Using the second option allows to display the elements in more detail. Figure 2.9 shows a package named *Generation* that contains two classes `HydroTurbine` and `PhotovoltaicModule`. Below this illustration, the inclusion of packages using another presentation is shown: In fact this means, as already described, that the packages `Distributed` and `Central` are contained in the package `Generation`.

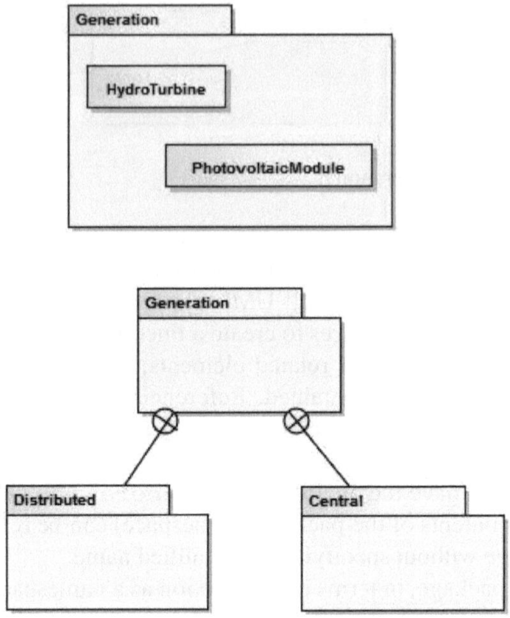

Fig. 2.9 Different Possibilities to Show Package Inclusion

2.1.4.1 Relationships between Packages

Beyond showing the contents of a package, package diagrams allow the clarify the relationships between packages. To depict the relationships between packages as for instance elements used from another package, the already introduced dependencies can be used. There are also methods to graphically present the already mentioned import option for packages using relationships. The package import is shown through a dashed line with an arrow at the end, which is drawn from one package to the imported one (on the side of the arrow), just like a dependency. The line gets its special meaning with the usage of the «import» stereotype (see section 2.1.5) that it is labeled with. By default, the imported elements are again publicly visible. If this is not desired, the «access» stereotype can be used. The elements from other packages will then be available within the package but only with private visibility.

An example of some dependencies can be found in figure 2.10. Here, the package `EnergyResources` makes the package `DistributedEnergyResources` publicly available in its context («import»). Both packages use elements defined in the package `Generation`.

For further information regarding package diagrams and its applications, please refer to [99].

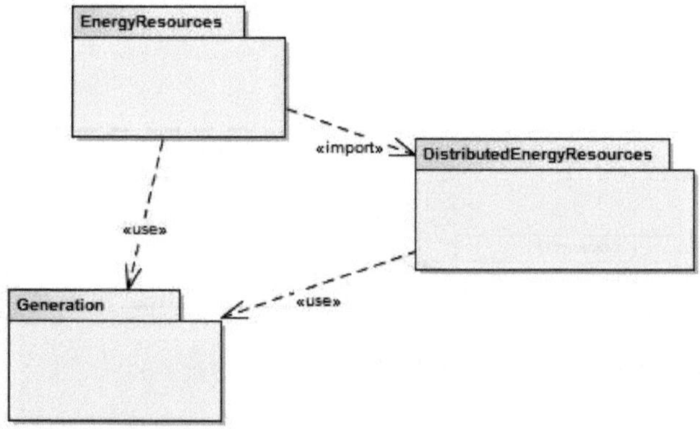

Fig. 2.10 Packages and Relations

2.1.5 Extending UML: Stereotypes and Profiles

In order to model specific things that for example are unique to a application area, it can be a better way to model these aspects with a specific "vocabulary" than with the generic elements the UML provides. If it is planned to extend the elements that can be used to model things, it can be described as extending the vocabulary that is allowed in UML – or more precise to change the meta-model. Extending the UML meta-model is possible with a mechanism called stereotypes. Stereotypes provide a simple mechanism to differentiate the meaning of an element, that can make a model in a specific context easier to understand.

Model elements (for example classes, attributes or relationships) can have zero or more stereotypes which are usually single words that clarify the role of an element. These words are enclosed in guillemots as e.g. «import» that could be assigned to a dependency relationship to clarify the kind of dependency in which one package imports another or «Asset» to tag a class on the first sight as an asset in a specific context.

In fact, a stereotype is a special class that is tagged with «stereotype» and defines properties that each element the stereotype is applied to should define. This is illustrated in figure 2.11: Having stereotyped elements allows tools to deal with them in a special way, as they might perhaps show another icon for classes with the stereotype «Asset». Moreover, this can become relevant in later phases when the model is used for code generation, where this information can be incorporated.

In practice, multiple stereotypes can be defined for the purpose of describing a specific application area. There can be restrictions defined using further features based on the UML (with OCL) and enriched with so-called tagged values that describe the model in more detail, what can be relevant for later usage and especially code generation. This set of UML additions that is used to describe a specific application area can, in order to be reused, aggregated as a UML Profile that can be used by tools. As it is not the focus of this book and the CIM does not make use of this, please refer to the technical literature for more details.

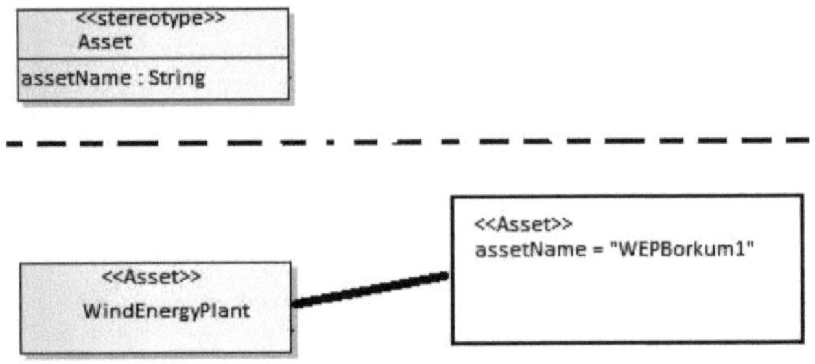

Fig. 2.11 UML Stereotypes and Their Application

The information provided up to this point should allow you to read the CIM UML model and to understand the syntactic elements used within the CIM. The CIM is a data model and so heavily makes use of classes and attributes to represent the information while other elements as operations can not be found.

Having the CIM described in UML finally allows its computer-aided use. The basic aspects you will be confronted with at that point will be files in which the structure and data is to be captured. XML is the format of choice for that and will be introduced in the following section.

2.2 XML Basics

UML as a formal language is mainly associated with its graphical representation and the concepts are widely understood in various domains. The implementation of what is modeled using the UML is not bound to a specific implementation language.

XML provides a convenient way to describe structured data in a textual representation. In general, text files are not platform or implementation specific and so XML is. Due to its structured approach, XML files can be read and interpreted by machines as well as by humans. The files can be exchanged independent of the platform and so they are for example being used for data transport between different computer systems across networks. With XML as a basic technology, it is also possible to exchange instance data that is defined according to the structure of the CIM which is described in the UML model.

The language grammar and its vocabulary is defined by the World Wide Web Consortium (W3C) as a technical recommendation in version 1.1 [124] and is (in fact) a subset of the Standard Generalized Markup Language (SGML) [78]. XML originated in 1996 to be an exchange format that is usable over the Internet and support a variety of applications. The intention XML was developed with was also to allow to write software that can process documents that are also readable by humans. It was designed to have documents that also shall be easy to create but also formal and precise. The basic structure of XML will be explained in the following, but just to the extent necessary to understand most of the documents – special details can always be found in the specification [124].

The text used in XML documents basically consists of character data and markup. The markup serves as a structuring for the content of the document. In [124], markup it is described as follows:

> Markup takes the form of *start-tags*, *end-tags*, *empty-element tags*, *entity references*, *character references*, *comments*, *CDATA section delimiters*, *document type declarations*, *processing instructions*, *XML declarations*, *text declarations*, and any white space that is at the top level of the document entity (that is, outside the document element and not inside any other markup).

These kinds of markup are defined between angle brackets (< and >) and it is defined that those characters as well as the ampersand character (&) "*must not appear* in their literal form, except when used as markup delimiters, or within a comment, a processing instruction, or a CDATA section".

Moreover it is defined, that all text that is not markup is the *character data* of the document. The following text describes the structure of XML documents in more detail.

2.2.1 Constituent Parts of the XML

Elements

The XML can be logically structured with elements. Each document contains one or more elements that must always have a name and are delimited by start-tag (`<tag-name>`) and end-tags (`</tag-name>`). The names of the start- and end-tag have to be exactly equal in case and can begin with a letter, colon or underscore followed by various characters except the space character that delimits the name. Although it is possible, it is recommended not to use colons or underscores at the beginning of a name.

Between these tags, content can be placed. Content comprises character data, further elements, processing instructions or comments.

If no content of this kind is required, a combined version of these tags – called empty element tag – can be used, see listing 2.1.

Listing 2.1 Example for XML Elements

```
<element>
    Content
    <emptyelement />
</element>
```

Attributes

Additionally, start-tags can have one or more attributes that define key-value pairs. An attribute has a name (key) that has to be unique for an element. The name of the attribute is followed by an equals sign and the value is specified between two quotes. The attribute value can contain virtually any character data (see [124] for details) but does not contain any further tags.

Attributes are separated from the element name, that starts right after the opening left angle bracket, by a space character. A space character is also used to separate further attributes. Listing 2.1 illustrates how attributes are put into an element.

Listing 2.2 Example for XML Elements with Attributes

```
<element attributename="value">
    Content

    <emptyelement attributename="value1" anotherattribute="value2" />
</element>
```

Attributes can further be typed to restrict their values. The definitions and implications of these restrictions will be described in more detail in section 2.2.2.

2.2.1.1 Character Data and Other Markup Elements

Usually all characters will be parsed and interpreted. To also use special characters like angle brackets (<, >) that are normally disallowed or for example Greek letters as content, XML defines a way to include them by reference.

Character references and entity references are introduced to overcome this problem: "A character reference refers to a specific character in the ISO/IEC 10646 character set, for example one not directly accessible from available input devices. An entity reference refers to the content of a named entity."

References in general start with an ampersand symbol, followed by their identifier and end with a semi-colon. The identifier of character references starts with a hash symbol that is followed by the Unicode number (in decimal or hexadecimal notation) of the character to be shown (e.g. `µ` or `µ` to refer to the Greek character μ). An entity reference identifier usually consists of characters. There are some inbuilt entity references but it is also possible to introduce new named entities (see section 2.2.2). Inbuilt entity references are:

- `&` for an ampersand symbol (&),
- `<` for an left angle bracket (<),
- `>` for an right angle bracket (>),
- `'` for an apostrophe (') and
- `"` for an quotation mark (")

Character Data (CDATA) Sections

Another way to include specific character data without interpretation by the XML parser, are character data (CDATA) sections. These sections "may occur anywhere character data may occur; they are used to escape blocks of text containing characters which would otherwise be recognized as markup" [124]. The content of CDATA sections is placed between `<![CDATA[` and `]]>` – everything lying inbetween will not be treated as markup.

Comments

Comments in XML can be expressed between the markup elements <! -- and -->. They may occur "anywhere in a document outside other markup (...) [and] may appear within the document type declaration at places allowed by the grammar" [124]. Comments do not belong to the document's character data and so applications that process XML documents must not retrieve their text contents.

Processing Instructions

As the name already indicates, processing instructions are to provide instructions to applications that process the document. Processing instructions are defined between <? and ?>. A processing instruction starts with a target application specification that is intended to process the instruction. This processing instruction target must be a name similar to the name definition for elements. However, it is forbidden to use the name `xml` (in any case) as it is reserved for a special processing instruction at the very beginning of the document (see section 2.2.1.2). The target can then be followed by instructions that can be expressed with characters.

All these content types shown can occur in XML documents – some restrictions regarding the overall document structure exist. The following section will explain how these elements can be used together.

2.2.1.2 Document Structure

An XML document basically begins with a *prolog*, exactly one (root) *element* and "miscellaneous" constructs. Listing 2.3 outlines an exemplary XML document that follows this structure.

Listing 2.3 Example XML Document

```
<?xml version="1.1"?>
<!-- This is a comment after the preceding prolog -->
<!DOCTYPE PowerPlants PUBLIC "-//OFFIS//DTD PPLANTS 1.0//EN" "http://www.
   offis.de/DTD/powerplants10.dtd">

<PowerPlants>
   <WindEnergyPlant id="1">
     <Rotors length="20">
       <Rotor no="1" yearOfManufacture="2011" />
       <Rotor no="2" yearOfManufacture="2010" />
       <Rotor no="3" yearOfManufacture="2011" />
     </Rotors>

     <?someApplication ignoreTurbineDefinition="no"?>
     <Turbine power="3">
       <Description>
           The turbine could be described with some text, that can be read
              by humans &amp; machines.
       </Description>
       <Extras>
          Everything <![CDATA[ inside this section will be quoted verbatim,
              even <tags> ]]>
       </Extras>
     </Turbine>
   </WindEnergyPlant>
   <!-- Further elements could follow below ... -->
</PowerPlant>
```

The prolog usually starts with an XML declaration which is actually a processing instruction stating at least the XML version used for this document. In the given example listing it is `<?xml version="1.1"?>`.

Comments, further processing instructions or whitespace characters may follow then. After that, a document type declaration that defines the structure of the document may follow (see line 3).

This line denotes that the `PowerPlant`-tag can be validated using the document type definition (see [124] for more details) that can be publicly found under the given URI. Alternatively, an XML schema file could be referenced to provide a link to a structural definition of the file. In contrast to the document type definition, XML schemas are referenced using special attributes at the root tag instead.

The prolog is then followed by the root element that can of course include further nested elements. Finally, additional miscellaneous constructs (comments, processing instructions or whitespace) can follow. If multiple XML documents are used and processed, it may occur that elements have the same name but refer to a different thing. To provide an option to make the used names for elements and attributes unique, XML Namespaces are defined.

XML Namespaces

XML Namespaces are defined in [125] and qualify a name using an Internationalized Resource Identifier (IRI). A namespace defines a scope for an element or attribute by prefixing their name with a shortcut to a defined namespace, followed by a colon.

Namespaces are declared using an attribute called `xmlns` that is usually placed at the root element – listing 2.4 shows an example for the declaration and usage of namespaces.

Listing 2.4 Example for the Declaration and Usage of Namespaces

```
<?xml version="1.1"?>
<PowerPlants xmlns="http://www.offis.de/pplants"
             xmlns:cb="http://www.offis.de/cim-book">
<!--
  The MicroCHP-element is in the default namespace http://www.offis.de/
    pplants
  The explanation-attribute's namespace is http://www.offis.de/cim-book
-->
 <MicroCHP id="2" cb:explanation="Combined Heat and Power">
     <cb:example>
            This example-element's namespace is http://www.offis.de/cim-book
            .
            It is not part of the default namespace.
     </cb:example>
        <!-- ... -->
     </MicroCHP>
</PowerPlant>
```

The default namespace, that is used if no explicit namespace name is put in front of a element or attribute name, is defined using the `xmlns` attribute in line 2. An example for a prefixed namespace definition is given in line 3, where the namespace `cb` is defined. The identifier which is given as a namespace must be a syntactically valid IRI, but it must not exist. – it is rather used as a unique identifier.

The application of namespaces is illustrated in line 8 of the listing, where the namespace `http://www.offis.de/cim-book` with the prefix `cb` is applied to the `explanation` attribute. At this the element `MicroCHP` is in the default, unprefixed namespace `http://www.offis.de/pplant`.

In line 9, an element that also refers the namespace `http://www.offis.-de/cim-book` is defined using the prefix `cb`.

With namespaces, multiple markup vocabularies can be used in one document or combined from multiple documents. To define the document structure more rigorously, it is possible to define the intended structure of XML documents and so to specify constraints. This is described in the following section 2.2.2.

2.2.2 Structure and Constraints for Documents

In general, XML documents can be structured as required for the needs and contain almost any contents. As good as this flexibility is, as bad are its consequences for data exchange. To automatically interpret an XML document, the structure of the documents content with its used elements must be known. A possible approach is to describe this structure externally so that only documents following the specification will be accepted.

In fact, there are several approaches to solve this problem in different levels of granularity, for example XML Schema Definitions (XSD) and XML Document Type Definitions (DTD).

XML Schemas in particular, are defined in an XML document and allow to describe the structure of a class of documents as well as number of appropriate data types. The XML Schema specification made up of three recommendations: A primer [121] (part 0) that summarizes the other parts, one document that describes the structures [122] (part 1) and another part [123] (part 2), that describes data types which can be used to describe the element's and attribute's contents. With XML Schema it is for instance possible to describe element types, their order, their number of occurrences and namespaces or default values. Those definitions can further be references and so be reused to create more complex definitions or otherwise. Moreover, it provides data types as e.g. boolean values, strings, numbers or enumerations. These types can be specifically constrained to the needs, for example in length or range. A simple example for an XML schema can be found in listing 2.5.

Listing 2.5 Example for the Declaration and Usage of Namespaces

```
1 <xsd:schema xmlns:xsd="http://www.w3.org/2001/XMLSchema">
2 <xsd:complexType name="PowerPlant">
3   <xsd:sequence>
4     <xsd:element name="name" type="xsd:string"/>
5     <xsd:element name="power" type="xsd:decimal"/>
6     <xsd:element name="description" type="xsd:string" minOccurs="1"/>
7   </xsd:sequence>
8   <xsd:attribute name="id" type="xsd:ID" use="required"/>
9 </xsd:complexType>
10 </xsd:schema>
```

The example defines XML documents that start with the `<PowerPlant>` as root element (line 2), that has a required attribute `id` (line 8). This attribute is of the special type `ID`, which actually means that it must be unique within a whole document. The root element can contain the elements `name` and `power` and must contain a `description` element.

The XML Metadata Interchange (XMI) is standard developed by the OMG for exchanging metadata information using XML. It can be used for any metadata whose meta-model can be expressed in Meta-Object Facility (MOF). The most common use of XMI is as an interchange format for UML models, although it can be used for serialization of models of other languages, also.

Rule Conformity of Documents

Given a document definition, an XML document can be checked for its correctness. The XML specification [124] defines two levels of correctness, a document can be checked for well-formedness and validity.

An XML document is *well-formed* when it conforms the respective rules defined in [124]. This includes for example that the document has exactly one root element which is the outermost element, that every element has a start- and end-tag, the elements with their start- and end-tags are nested properly within each other and an element does not have multiple attributes with the same name.

To be *valid*, an XML document must be well-formed and have an associated document type declaration (defined with an Document Type Definition (DTD) or XML-Schema) to which it complies to. By the usage of this check, major mistakes can easily be recognized at this level. XML parsers can check the documents whether they conform to the respective rules and will usually stop processing if this check fails.

2.3 RDF Basics

The previous section introduced XML as an enabling technology to build upon, as it is well structured and machine readable. But one major disadvantage is, that its tags are readable and - in most cases - are understandable for humans, but there is no link to what they really mean except their name. Their name usually assumes a specific thing that is associated with it and so has a meaning to humans. This lack of semantic is adressed by RDF that focuses on the description of resources.

It is defined by a set of W3C recommendations[1], as e.g. [117], [118], [119] or [120], and is intended to serve as a standard model for meta data interchange on the web. Its approach and application area goes beyond the web and beyond just the syntactic description of data, as it is for instance possible with XML and XML schema. RDF provides features to semantically describe information and so to facilitate the integration of multiple information sources. That means information can be exchanged without loosing its meaning and since RDF provides a common framework, its advantage is that it is processable by machines by being still readable by humans. RDF in the context of the CIM is for example used to serialize network topology data.

Basically, RDF describes resources, identified by Uniform Resource Identifiers (URIs), and sets them in relation to other resources. This is realized using a triple made up of a subject, a predicate and an object. The subject describes the resource about which a statement shall be expressed, the predicate refers to a certain property of an object and the object serves as an argument for the predicate.

This emerging structure can be represented as a directed graph having the resources (subject and object) as nodes and the predicate as (labeled) edge. Resources that are subject or object of a triple are symbolized by an ellipse and literals, representing data values, are symbolized by rectangles. Literals can never be origin of edges in an RDF graph. Figure 2.12 shows an example for an RDF graph making a statement about a wind energy plant.

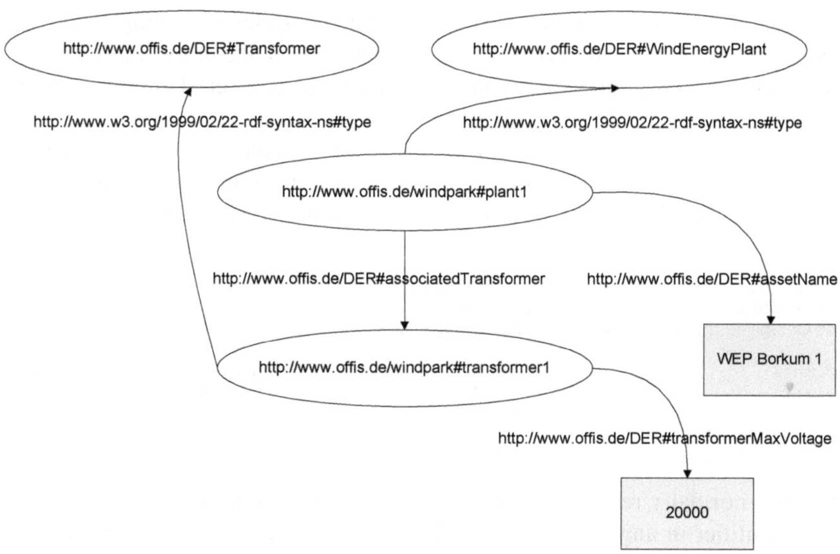

Fig. 2.12 Example RDF Graph

[1] http://www.w3.org/RDF

The graph provides information about an individual wind energy plant `http://www.offis.de/windpark#plant1` that is typed as one of its kind using the RDF type predicate `http://www.w3.org/1999/02/22-rdf-syntax-ns#type`. In concrete terms its type is identified by `http://www.offis.de/DER#WindEnergyPlant`.

This specific plant has an asset name, as defined by `http://www.offis.de/DER#assetName`, that is `WEP Borkum 1`. Moreover, the wind energy plant has a concrete associated transformer `http://www.offis.de/DER#associatedTransformer` with the resource URI `http://www.offis.de/windpark#transformer1`.

This transformer has further properties `http://www.offis.de/DER#transformerMaxVoltage`, that is a maximum voltage of 20.000 (volts).

This exemplary graph illustrates multiple concepts, which can be expressed with RDF, that are:

- Specific classes of things, e.g. a `WindEnergyPlant` with the identifier `http://www.offis.de/DER#WindEnergyPlant`
- Properties of those classes, `http://www.offis.de/DER#assetName` e.g. asset name to be identified with
- Instances of these classes (individuals), as e.g. a specific wind energy plant, identified by `http://www.offis.de/windpark#plant1`.
- Values of class properties, e.g. `WEP Borkum 1` as an asset name property instance

To represent this graph as structure in a textual form, the triples consisting of subject, predicate and object can be described. As we already saw in the graph, most of these are represented by URIs and so can be referenced. Several serializations of RDF graphs exist, like N3, Turtle and XML/RDF.

An descriptive notation that expresses this tuple structure quite well is Turtle. Due to its terseness, it is easy to read and can often be found in literature. The graph above can be represented according to this notation, as shown in the following listing:

```
1 @prefix rdf: <http://www.w3.org/1999/02/22-rdf-syntax-ns#> .
2 @prefix der: <http://www.offis.de/DER#> .
3 @prefix wp:  <http://www.offis.de/windpark#> .
4
5 wp:plant1                 rdf:type                        der:WindEnergyPlant .
6 wp:plant1                 der:assetName                   "WEP Borkum 1" .
7 wp:plant1          der:associatedTransformer   wp:transformer1 .
8 wp:transformer1    rdf:type                    der:Transformer .
9 wp:transformer1    der:transformerMaxVoltage   "20000" .
```

The notation normally requires to define resources and predicates with their fully qualified identifier in angle brackets. As this introduces a large overhead, prefixes can be defined (line 1-3).

Typically, the RDF namespace has the prefix `rdf` and is referenced using the URI `http://www.w3.org/1999/02/22-rdf-syntax-ns#` (line 1). What follows is a collection of subject, predicate and object tuples describing the graph (lines 4-8). In line 4, a "windpark"-individual of the type `WindEnergyPlant` is defined.

This individual, referenced as `plant1` (subject), has the `assetName` (predicate) `WEP Borkum 1` (object) (line 5). Turtle furthermore provides even a shorter notation to describe reoccurring resources that will not be described in detail here.

However, the most common serialization that can be found in practice, is an RDF/XML serialization. Having the graph described in XML, simplifies the data exchange as it is a widespread format and also allows to augment XML based documents as e.g. HyperText Markup Language (HTML). The following listing describes still the same example with the XML syntax.

```
1  <?xml version="1.0"?>
2  <rdf:RDF xmlns:rdf="http://www.w3.org/1999/02/22-rdf-syntax-ns#"
3           xmlns:der="http://www.offis.de/DER#"
4           xmlns:base="http://www.offis.de/windpark#">
5
6      <!-- Define a resource of the type WindEnergyPlant -->
7    <der:WindEnergyPlant rdf:ID="plant1">
8      <!-- Assign the property assetName -->
9      <der:assetName>WEP Borkum 1</der:assetName>
10     <!-- Reference another resource,
11     identified by http://www.offis.de/windpark#transformer1 -->
12     <der:associatedTransformer rdf:resource="transformer1"/>
13   </der:WindEnergyPlant>
14
15   <der:Transformer rdf:ID="transformer1">
16     <der:transformerMaxVoltage>20000</der:transformerMaxVoltage>
17   </der:Transformer>
18 </rdf:RDF>
```

The XML document's root element is in this case `RDF` within the `rdf` namespace (line 2). Furthermore the relevant namespaces are declared at the beginning. The `base` namespace `http://www.offis.de/windpark#` in particular defines the base URI that is used to resolve relative RDF URI references.

In line 7 the `WindEnergyPlant` (using the namespace name `der`) individual is declared and has the relative RDF resource identifier `plant1` that will be resolved using the URI given in the `base` namespace.

A literal with the value `WEP Borkum 1` (object) is defined in line 9 as the `assetName` (predicate) of this individual (subject). Finally, the resource with the relative identifier `transformer1` is referenced as `associatedTransformer` in line 11. In addition to that, RDF/XML serialization provides further language constructs which can be found in literature and the corresponding specifications.

On top of that, *RDF Schema* (RDFS) can be used to define a vocabulary to define particular application areas and so formalize simple ontologies. It extends RDF's range with classes, properties, instances and subclasses as basic modeling primitives for example. Those RDF Schemas relate to RDF documents as for instance XML schema does for XML. RDF Schemas themselves are defined in RDF – a simple example is provided in the following listing.

```
1 <?xml version="1.0"?>
2 <rdf:RDF
3   xmlns:rdf="http://www.w3.org/1999/02/22-rdf-syntax-ns#"
4   xmlns:rdfs="http://www.w3.org/2000/01/rdf-schema#"
5   xml:base="http://www.offis.de/DER#">
6
7   <rdfs:Class rdf:ID="DistributedEnergyResource"/>
8
9     <rdfs:Class rdf:ID="WindEnergyPlant">
10    <rdfs:subClassOf rdf:resource="#DistributedEnergyResource"/>
11    </rdfs:Class>
12
13    <rdfs:Class rdf:ID="PhotovoltaicPowerPlant">
14    <rdfs:subClassOf rdf:resource="#DistributedEnergyResource"/>
15    </rdfs:Class>
16
17    <rdfs:Class rdf:ID="Transformer"/>
18
19    <!-- Further definitions -->
20 </rdf:RDF>
```

In this example, some types (classes) of distributed energy resources are defined, that were already used in the earlier example. At first, the DistributedEnergyResources class is defined, using the respective `rdf:ID` attribute (line 7). Some specialized class definitions, that are actually subclasses of the of a distributed energy resources, follow. This subclassing can be best described as an "is-a" relation between the involved entities. It is expressed with the `rdfs:subClassOf` tag in lines 10 and 14. An instance of this schema would look quite similar to the previous examples shown, except it does not define any properties.

RDF Schema (RDFS) allows to define further restrictions that can not be treated in detail here. The RDFS specification [119], the RDF Primer [117] or advanced literature as e.g. [23] provides further information on that.

Moreover, additional technologies to express semantic connections exist. Another prominent one that builds on RDF and RDFS, is the Web Ontology Language (OWL) [127, 128] which is also specified by the W3C World Wide Web Consortium. It exceeds the expressiveness of those languages and is also often mentioned in the context of semantic data description.

Having such a well-defined and formal specification that is machine readable allows advanced applications that would not possible on unstructured and non formal defined data, as for instance to query on these data. One query language in this context is for instance SPARQL Protocol and RDF Query Language (SPARQL) for RDF. Additionally, it is possible to automatically reason about the data an infer relations between individuals.

2.4 Techniques for Integration

While systems and applications in the past were often operated separately, today business requires interactions between multiple systems and applications to operate effectively. To do so, a coupling of former separated and heterogeneous systems is necessary. This requires solutions to integrate those systems in a way their functionality is still available and can be adapted to changing need. The establishment of a common information model that is to be used throughout many applications and systems requires solutions to cope with different data sources from the various actors.

To allow the recombinations of different data sources and the establishment of new interfaces between those systems, syntactic and semantic interoperability is required. Different than in data or function integration, the implementation of the original systems is not affected by this enterprise application integration approach.

Usually, the integration will be realized through integration platforms that allow the implementation of required interfaces – Middleware is often the layer where this integration effort takes place.

Often times the middleware is message-based, meaning components exchange data defined in messages which are sent from one component to another. XML is then used for various purposes, like message description and interchange. By shifting the intelligence to interfaces in conjunction with intelligent routing, publish/subscribe and event mechanisms, it is possible to define performant systems spanning across multiple organizations and actors. In general this approach is labeled as EAI.

SOA goes one step further with the integration approach as well as with the organizational embedding, but also share the technological concepts of EAI. A SOA requires specific features according to the service paradigm from the applications to be coupled in order to allow for successful process integration. The smallest units in SOAs are services that provide a defined set of functionality, being so fine-grained that they provide units for reuse.

However, what exactly a service is and its level of granularity is in many cases defined different. The term service is often considered from a certain perspective from a particular stakeholder group, for instance, regarding the structuring of the business or the IT, as being stated by TOGAF [111]. Different approaches to describe SOAs and to classify their services are further mentioned in [100].

Services, both business and technical, are self-contained, have a contract assigned that specifies their functionality and how to access it, and produce predictable results. In contrast to the sizing of applications and their functionality, services are designed to be used with other services in terms of composition and orchestration. This level of granularity adds more flexibility to business processes, as they may defined and executed using the services. Services are characterized by loose coupling and will usually be provided in specific directories where they can be found by third parties or process engines. Technical services are mostly realized as Web Services using the WS* technology stack from W3C. The service localization is then realized with Universal Description, Discovery and Integration (UDDI) that provides standardized way to locate those services. Besides the possibility of direct coupling, the usage of a platform providing the required functionality to orchestrate and compose services is highly important and can be realized in SOA middleware.

The features that middleware for this application area usually offers can for example be data transformation, connection to data sources, automation technology, logging, reporting as well as filtering and transformation. Such complex middleware is often named ESB. A platform like this can serve as a focal point for data, but it can also become a bottleneck for the decentral arranged services. Therefore, it is beneficial to have a redundant middleware infrastructure that is scalable. In case a part of the IT infrastructure is not operated by a company itself but by another provider, the provided infrastructure becomes more and more abstract and blurred, meaning it appears as to be surrounded by a cloud.

By turning from a central IT to more decentralized systems in the energy sector, more efforts on system integration have to be spent. The mentioned integration paradigms are very valuable for Smart Grids, as they can be applied for the integration of decentralized systems, comprising producers, storage, consumers and other data sources. Here, the integration paradigms of EAI and SOA may be used for communication, automation as well as for secondary and primary IT. Internationally standardized solutions already exist to simplify this, like for instance the IEC 62357 SIA, which can be realized using a SOA, or the IEC 62541 OPC Unified Architecture as a SOA-based approach for data exchange. Nevertheless, there are still gaps that require harmonization between semantic and syntactic interfaces.

2.4.1 Outlook on Further Developments in Integration Technologies

Today several of the following basic technologies exist and are wide spread in some industries (e.g. in banking and e-commerce). Currently, the paradigm of the integration of heterogeneous systems using modern IT-integration technologies is characterized by a monolithic approach.

This results in highly integrated systems, connected with point-to-point connections. EAI technologies and solutions are used for the realization of those systems but there is further potential for optimization. In order to promote the integration, a wide establishment of SOAs is a first step. SOAs do not describe out-of-the-box solutions but rather a paradigm for the development of systems that fulfill individual requirements. Developers are provided with a structure that allows different, distributed systems to be combined. As an architectural paradigm, SOA is concerned with the management of business processes that are distributed over numerous of existing and new, heterogeneous systems. These systems are moreover often controlled by different actors [82]. Thus, SOA requirements fulfill the Smart Grid actors' requirements regarding the IT infrastructure, as they also show a high degree of distribution.

Having established this, puts the focus on syntactic interoperability (see figure 2.13). This means, that two or more systems are able to communicate with each other and to exchange information. The specification of data formats, data serializations and communication protocols is essential to achieve this goal. Standards like XML or Structured Query Language (SQL) with their dialects can act as enablers for syntactic interoperability. Syntactic Interoperability is the prerequisite for other forms of interoperability as for instance semantic interoperability [113].

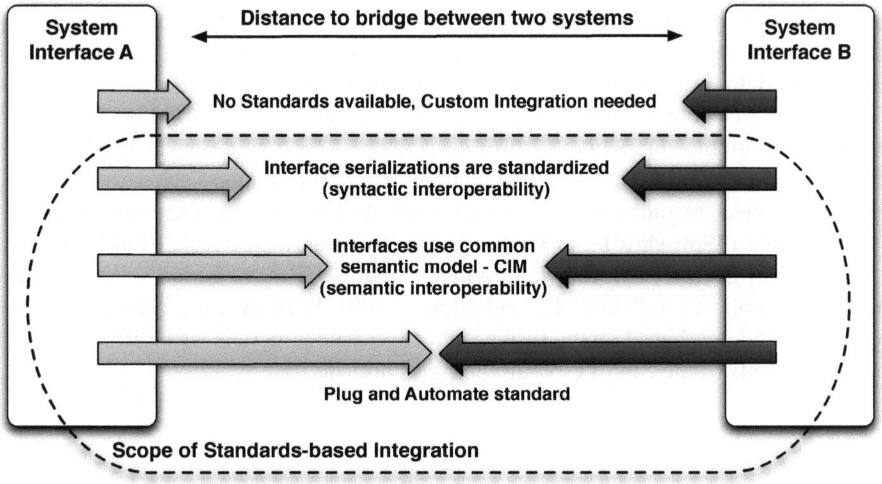

Fig. 2.13 Distance to integrate

After the level of syntactic interoperability is reached, appropriate reference architectures for the development of new systems can be defined. Reference architectures describe a certain type of domain specific architectures and moreover a large class of systems in an abstract manner. So they allow developers to obtain information about the design of specific classes of systems. Reference architectures are usually being developed by deriving the information from domain specific studies. All in all, reference architectures depict an ideal architecture with all functionalities that can be realized through concrete systems [109].

Semantic interoperability then decreases the integration distance in comparison to the mentioned reference architectures (see figure 2.13). It is defined as the capability to utilize the information exchanged to achieve results that are usable by all involved systems. In order to reach semantic interoperability, all involved systems have to agree upon a single, content-oriented reference model for information exchange. The contents of information exchange have to be defined clearly – the information sent is semantically the same what is interpreted by the receiver [113].

Finally, the integration across clouds in the energy domain is what can be seen as the last step as of today. Similar as with the term Smart Grid, there are different definitions of what cloud computing is. A widely accepted definition is given and continuously revised by the NIST [90].

Cloud Computing according to the NIST is defined as a model that allows ubiquitous, comfortable on-demand network access to a common pool of configurable electronic data processing resources like servers, storage, applications and services. Those resources can in addition to that be used and provided quickly, that means with minimum management efforts or service provider interactions. Cloud computing comprises besides five essential characteristics (on-demand self-service, broad network access, resource pooling, rapid elasticity and measured service) three service models ("Software as a Service", "Platform as a Service", "Public Cloud" and "Hybrid Cloud") and four deployment models ("Private Cloud", "Community Cloud", "Public Cloud" and "Hybrid Cloud") [90]. With this, the prerequisites for business models – based on the service and deployment models – for different actors as participants of the Smart Grid are being established.

Chapter 3
The IEC Common Information Model

The CIM is standardized within three different IEC standard series, namely IEC 61970, IEC 61968 and IEC 62325. Each of them has a different background and covers different use cases. However, all of them consist of many sub parts which are under continuous development lead by distinct working groups. Since they all rely on the same data model strong liaisons have been established and the CIM user group was founded as a central place for discussions and cooperations. The three standard series have the following focus:

- **IEC 61970:** deals with "Energy Management System Application Program Interfaces (EMS-API)" and is maintained by IEC TC 57 WG 13. The series consists of several parts being based on the research report RP-3654-1 from the EPRI CC-CAPI project. Against the background of providing a comprehensive integration framework the standard series specifies various components. Beside an extensive data model defined within IEC 61970-301, it contains CIS as well as GID. One of the pursued objectives is to develop a platform independent data model using technology independent services which is continuously improved. However, mappings to specific technologies are or will be specified like RDF, XML and OWL.
- **IEC 61968:** is called "Application Integration at Electric Utilities - System Interfaces for Distribution Management" and contains (amongst others) the IRM for the CIM. In contrast to the CIS/GID where generic interfaces are defined within IEC 61970, the IRM specifies detailed use cases for coupling two systems and the according exchanged XML messages. Thereby, the focus is not on the CIM objects primary needed for grid operations, but on secondary objects like billing and network extension. The CIM base data model defined in IEC 61970-301 is extended with further objects within IEC 61968-11. The addressed interfaces cover message exchange for network operations, records and asset management, operational planning and optimization, maintenance and construction, network extension planning, customer support, and meter reading and control. The standard series is maintained by the IEC TC 57 WG 14.
- **IEC 62325:** is maintained by IEC TC 57 WG 16 and includes a set of standards describing a "framework for energy market communications". The scope is to

M. Uslar et al.: The Common Information Model CIM, POWSYS, pp. 75–106.
springerlink.com

develop standards for electricity market communications on the premise to use the IEC TC 57 CIM. Communications between market participants and market operator as well as communication among market operators are covered by the standard series. Two styles of market are distinguished. On the one hand, "European style markets" and on the other hand, "US style markets". The applied methodology comprises the following steps: documentation of data exchange requirements (use cases, workflows, etc.), mapping to existing CIM UML model, extensions to form new revision of CIM UML model, generation of contextual models, generation of a data exchange profile, generation of data exchange formats and use of Open Source tools like CIMtool (see chapter 6.2.3).

Figure 3.1 gives an overview on the standards which are currently parts of the three standard series. Those standards were considered whose status was presented by the according model manager at the last meetings. Each of the standards will be introduced within this chapter considering the current status (as time of writing in September 2011).

3.1 General Information

This chapter describes four sub standards, two from each standard series, the scope of which is to give an introduction to the according series of standards and to define the most significant terms. The order of the parts described is based on their function, which means that not all the parts form 61970 or 61968 are in one section in an canonical order but data models, general information, business functions and informations exchange are the categories used for grouping.

3.1.1 IEC 61968-1: Application Integration at Electric Utilities - System Interfaces for Distribution Management - Part 1: Interface Architecture and General Requirements

In general, the standard series is intended to foster inter-application integration of those distributed software application systems, supporting the management of utility electrical distribution networks, so the focus is on DMS. Thereby, the systems being integrated are supported by dissimilar runtime environments and are used to exchange data triggered by certain events. They are loosely coupled and heterogeneous in terms of languages, operating systems, protocols and management tools. IEC 61968 is used complementary to existing utility data warehouses, database gateways and operational stores, as it is implemented with middleware services connecting applications.

The first sub part of the standard series [62] introduces the interface architecture and general requirements. Figure 3.2 shows different grouped systems being part of a DMS and connected via interfaces to an application integration infrastructure. Those interfaces are defined by identified and established requirements in this sub part and they are the basis for the following parts of the standard series, as one

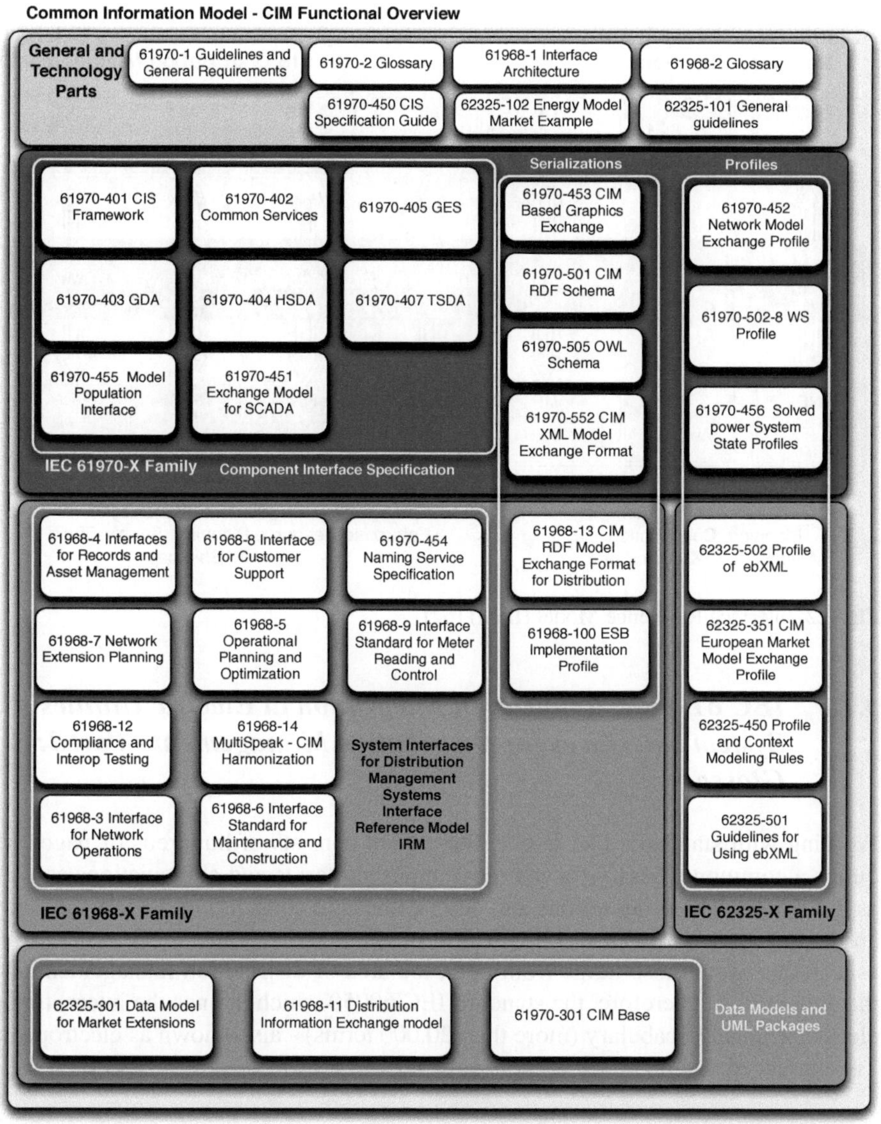

Fig. 3.1 Overview on the CIM Parts for 61968, 61970, 62325

sub part covers one interface. The interfaces are discussed technology independent, which mean that technologies and methods being used to implement the interfaces are out of the scope of the standards.

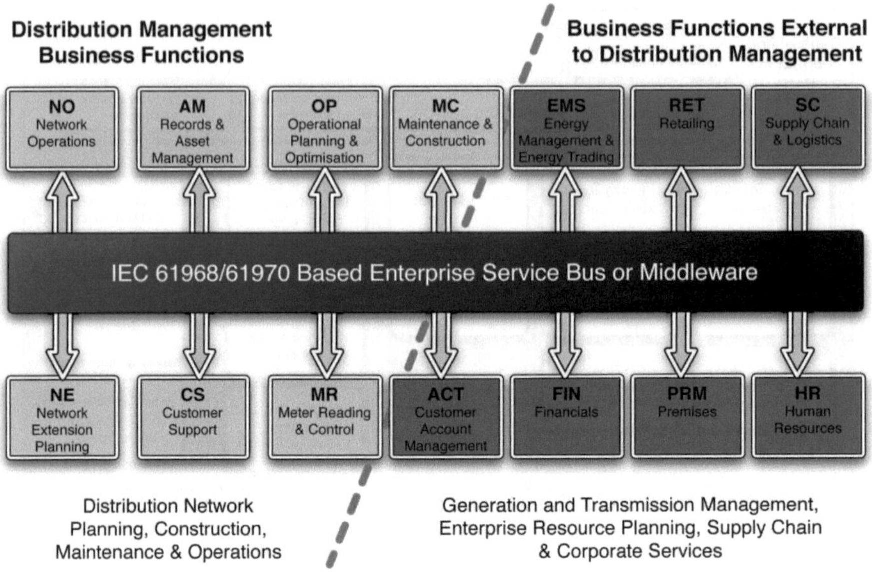

Fig. 3.2 Interface Reference Model (IRM) [62]

3.1.2 *IEC 61968-2 Application Integration at Electric Utilities - System Interfaces for Distribution Management - Part 2: Glossary*

Working with standards like IEC 61968 which consists of hundreds of pages requires a common glossary for the most important terms and abbreviations as well as definitions. Those definitions and descriptions are considering the terms as they are used in the context of the IEC 61968 standard series. Part 2 [70] of the series especially covers those terms from both, the area of distribution management and the area of ICT. Therefore, the standard IEC 60050 which defines the international electrotechnical vocabulary (more than 20.000 terms) - also known as electropedia - is referenced.

3.1.3 *IEC 61970-1 Energy Management System Application Program Interface (EMS-API)- Part 1: Guidelines and General Requirements*

The IEC 61970 standard series mainly deals with the definition of an API for EMS. As mentioned in chapter 1.4, the work is based on EPRI's CCAPI research project. The main objective of IEC 61970 is to provide a standard series including guidelines which support both, the integration of multi-vendor application systems for control center and information exchange with systems being outside of the control center environment like other transmission systems, distribution systems and generation systems. Thereby new systems are considered as well as existing legacy systems.

Part 1 [34] of the standard series includes basic guidelines and general infrastructure capabilities which are needed to apply the interface standards. Beside typical integration scenarios also types of applications which should be integrated were considered. Furthermore, a reference model is introduced covering the application of the other parts of the series. It is based on a component architecture comprising interfaces for information exchange allocated to the single sub standards.

As the integration infrastructure is not part of the standard, only common technologies are described which facilitate information exchange via the specified interfaces of the standard series. Moreover, the specifications of interfaces are limited to those being externally visible and improve the interoperability between products from different vendors. This is the reason why the standards do not include customized implementations and products nor restricting the representation of information within single applications.

3.1.4 IEC 61970-2 Energy Management System Application Program Interface (EMS-API) - Part 2: Glossary

Part 2 [28] of the IEC 61970 series is structured analogous to part 2 of the IEC 61968 series. Specific terms which need interpretations and abbreviations are defined as how they are used in the context of the IEC 61970 standards.

3.1.5 IEC 62325-101 Framework for Energy Market Communications - Part 101: General Guidelines

Part 101 [30] of the IEC 62325 standard provides technology independent guidelines and requirements for e-business in energy markets, which are based on Internet technologies. It supports the communication aspect of e-business applications in deregulated energy markets with focus on system operator including interfaces between market participants. However, the process real-time communication of energy systems is not in the scope of this standard.

An important note is that this standard series will not specify any content regarding processes, collaborations, transactions, messages or core components because of the fact that markets still vary. Only generally modeling tools and methods are described.

3.1.6 IEC 62325-102 Framework for Energy Market Communications - Part 102: Energy Market Model Example

This part [31] of IEC 62325 provides an example for the use of the methods and tools of the 101 part described in chapter 3.1.5. It mainly deals with the UN/CEFACT modeling methodology modeling of the energy market. This part can only be seen as informative example. The main purpose is to show how to apply the modeling methodology. As a result of the example, a business process and information model is generated.

3.2 Data Models

One of the most important parts of the CIM comprising the three standard series is the extensive data model. The base model is defined in IEC 61970-301 and extended with objects used for distribution management by IEC 61968-11 and with objects for market communication by IEC 62325-301. SPARX Enterprise Architect is the tool of choice maintaining the combined data model in the UML format. The development of the model is a steady process but for the annual releases, the model is frozen to publish a new release. If a new edition of the three standards which include the data model should be published a tool is used to generate that part of their contents describing the according data model. Since the combined data model is the core of the CIM, almost all other parts of the three standard series rely on the model. Thus, it is used to model the grid topologies or to specify the content of exchanged messages.

3.2.1 IEC 61970-301 Energy Management System Application Program Interface (EMS-API) - Part 301: Common Information Model (CIM) Base

Part 301 [43] is the largest part of the standard series. The predominant content are descriptions of the UML objects of the CIM. Figure 3.3 shows an overview on the structure of the model's packages which group the underlying classes by their functionalities and concepts. The classes, again, have attributes and are connected by associations to express dependencies and to enable modeling complex objects. The defined object model represents a logical view on physical aspects of EMS including SCADA. This standard should be understood as an instrument allowing integration in each domain needing a common model in terms of electricity. Within those domains, the model should be used to foster interoperability and compatibility between applications and systems independent from their implementations.

Table 3.1 [116] gives an overview on the quantity of the CIM objects comparing two former versions of the electronic data model. The increasing number of CIM objects emphasizes the development of the model. The shown amount of classes, attributes and associations also include those objects defined in the IEC 61968-11.

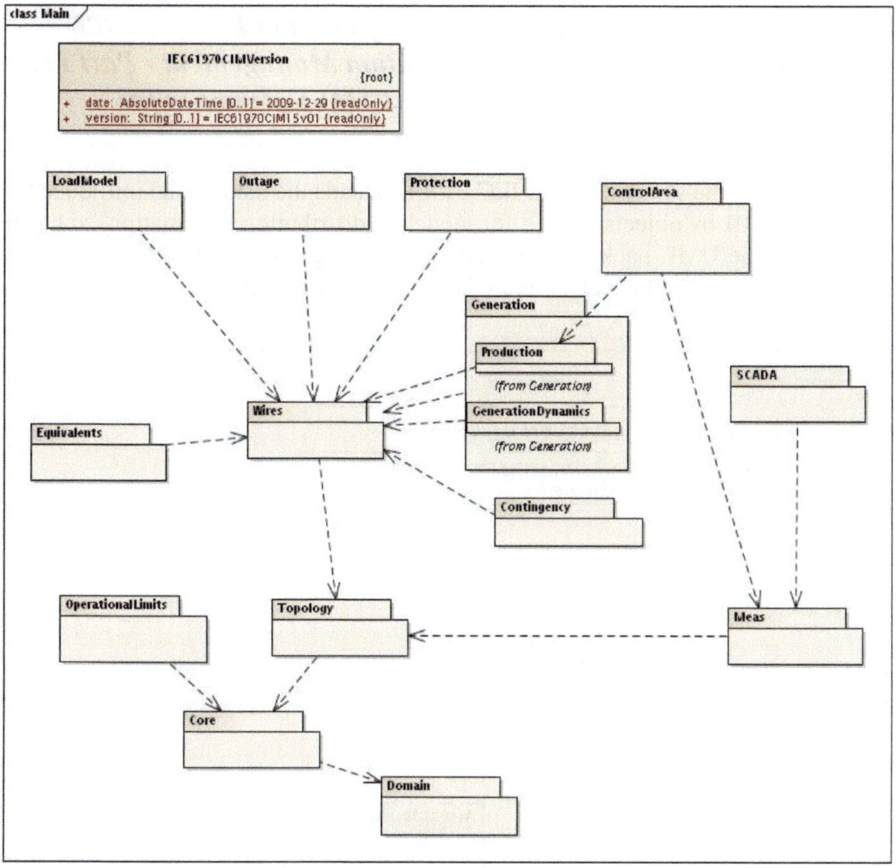

Fig. 3.3 IEC 61970-301 CIM base data model [43]

Table 3.1 Quantitative comparison of CIM releases 11 and 13

	Release 11	**Release 13**
Packages	53	45
Classes	≈ 800	≈ 900
Associations	≈ 780	≈ 870
Native attributes	≈ 2000	≈ 2650

3.2.2 *IEC 61968-11 Application Integration at Electric Utilities - System Interfaces for Distribution Management - Part 11: Common Information Model (CIM) Extensions for Distribution*

Part 11 [56] of the standard series IEC 61968 extends the data model introduced in IEC 61970-301 by objects needed for modeling distribution information exchange purposes. Nine UML-packages were added including additional classes, attributes and associations. The logical structure is depicted in figure 3.4.

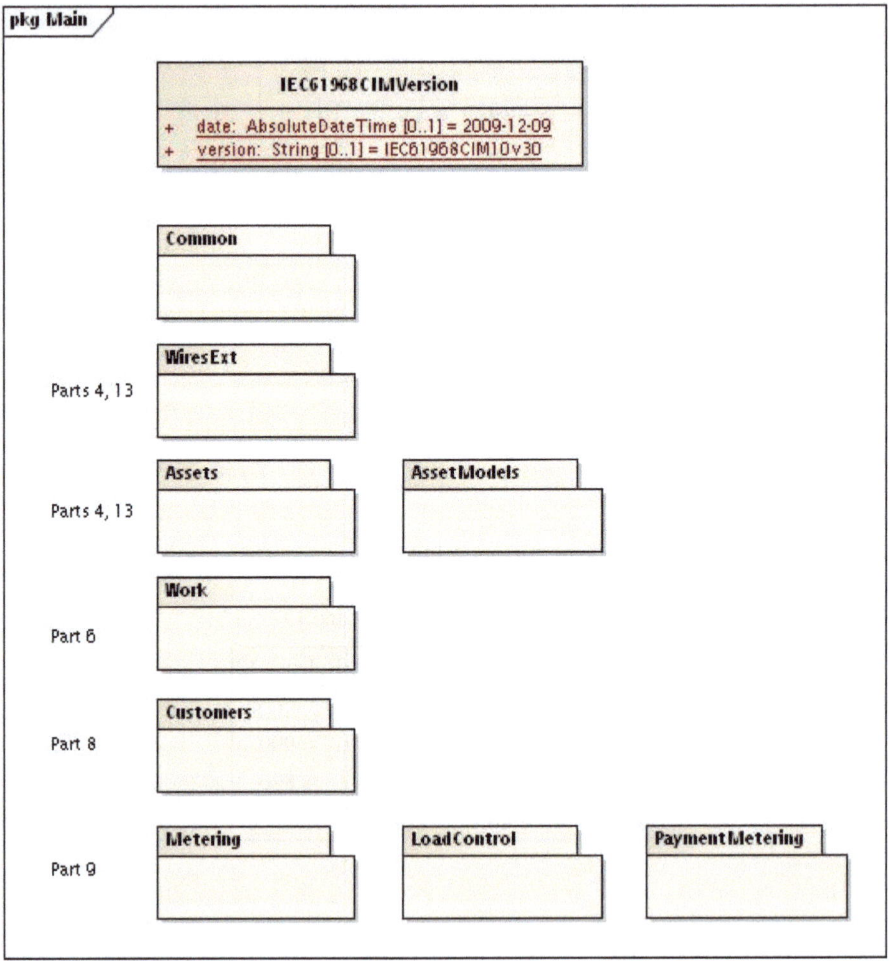

Fig. 3.4 IEC 61968-11 CIM data model [56]

The *Common* package includes information which are used to describe core functions of distribution management in general. The package *WireExt* contains the information classes which extend the *Wires* package from IEC 61970-301 with power system resources required for distribution network modeling, including unbalanced networks. Classes supporting asset management applications with specialized classes for asset-level models for objects are included in the *Asset* package. Whereas, *AsselModels* is an extension to that package containing classes for asset management and different network and work planning applications with specialized documentation classes describing assets of a particular product model made by a manufacturer. There are typically many instances of an asset associated with a single asset model. The packages *Work* and *Customers* consist of classes enabling work management and network extension planning applications respectively customer billing applications. The classes described in the *Metering* package are generally associated with the point where a service is delivered to the customer and therefore, support end device applications with specialized classes for metering equipment and remote reading functions. This packages is extended by the *LoadControl* package by classes considering specialized applications such as demand side management using load control equipment. The last package *PaymentMetering* is also an extension to the *Metering* package. Its classes deal with specialized applications such as prepayment metering. They are generally associated with the collection and control of revenue from the customer for a delivered service.

3.2.3 IEC 62325-301

The IEC 62325-301 draft version provides the extensions to the existing 61968 SIDM and 61970 EMS models. The 62325-301 and its sub-packages depend on the existing data model and extend them. Mainly, the IEC 62325 model consists of three packages, MarketsCommon, MarketOperations and MarketManagement.

The first package, the MarketManagement package provides a vast number of classes which should be used alongside the Common Market Model to generate a profile. This profile should be used iff the electricity market is mainly based on regulated third party access (e.g. TSO have to allow any supplier non-discriminatory access to the transmission network and wholesale and retail market transaction are used to exchange energy.

The second main package, MarketOperations deals with the same purpose as the MarketManagement package, e.g. to create a profile alongside the existing CIM data model. For this package, the focus is on the US-style electricity market models. Those differ from the European-style markets and mainly focus on the following aspects. The markets are characterized by day ahead unit commitment my a market operator, both intraday and real-time balancing through some central dispatch and settlement based on locational marginal prices. Apart from the data model itself and the CIM extensions, the -301 part provides an introduction to the most typical systems used in US-Style electricity markets and their architecture using ESB at Market operators.

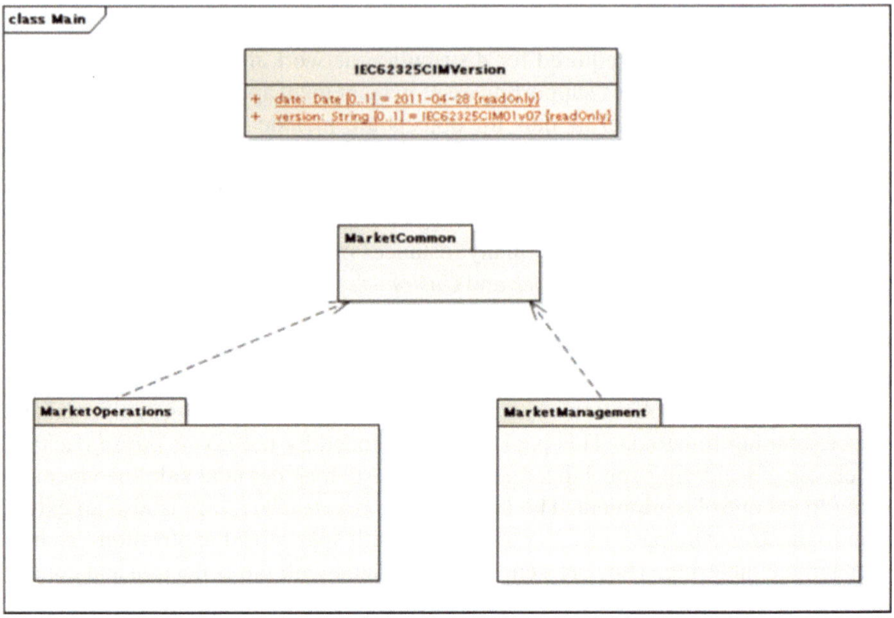

Fig. 3.5 IEC 62325-301 CIM data model

3.3 Business Functions

Parts 3 to 10 of IEC 61968 specify those interfaces describing the most important business processes from the reference model depicted in figure 3.2 and introduced in IEC 61968-1 (see section 3.1.1). Within the standard series, a distribution management system consists of many distributed application components. These components are needed by energy service provider to manage electricity distribution grids. Thereby, messages for the following processes are standardized:

- monitoring and control of equipment for power delivery
- management processes to ensure system reliability
- voltage management
- demand side management
- outage management
- work management
- automated mapping
- facilities management

3.3.1 IEC 61968-3 Application Integration at Electric Utilities - System Interfaces for Distribution Management - Part 3: Interface for Network Operations

Part 3 [27] defines an interface for network operations. An amount of message types is defined, which are used to cover the most business processes for operating networks. These message types are typically used for the following processes:

- data acquisition by external systems
- fault isolation
- fault restoration
- trouble management
- maintenance of the plant
- commissioning of the plant

3.3.2 IEC 61968-4 System Interfaces for Distribution Management - Part 4: Interface Standard for Records and Asset Management

This part [42] describes interfaces for records and asset management. Within the standard, only message types for records and asset management were covered by XML- schemes. Those messages comprise the areas of network extension planning, copying feeder or other network data between systems, network or diagram edits and asset inspection.

3.3.3 IEC 61968-5 Application Integration at Electric Utilities - System Interfaces for Distribution Management - Part 5: Operational Planning and Optimization

Part 5 is currently a working draft and postponed until experts from five countries are provided[1]. The standard should be based on revised IEC 61968-3 and will focus on interfaces for operational planning and optimization, applicable for the following use cases:

- carry out planned maintenance of High Voltage plant
- carry out planned maintenance with temporary equipment
- optimize voltage
- operational planning
- notify customer
- do field work under safe conditions
- fetch network data

[1] As presented in May 2011 at the WG 14 meeting in Prague.

3.3.4 IEC 61968-6 Application Integration at Electric Utilities - System Interfaces for Distribution Management - Part 6: Interface Standard for Maintenance and Construction

Similar to the aforementioned parts, part 6 [50] also defines an interface dealing with maintenance and construction. Analogously, the message contents with regard to the business processes were specified. Typical use cases are planned maintenance, unplanned maintenance, conditional maintenance, work management and new service requests.

3.3.5 IEC 61968-7 Application Integration at Electric Utilities - System Interfaces for Distribution Management - Part 7: Network Extension Planning

Similar to IEC 61968-5, IEC 61968-7 is still a working draft postponed until experts from five countries are provided[2]. The standard will cover interfaces and according use cases for network extensions planning.

3.3.6 IEC 61968-8 Application Integration at Electric Utilities - System Interfaces for Distribution Management - Part 8: Interface Standard for Customer Support

Another interface is addressed by part 8 [71]. In this case, messages for customer support are defined. Many business functions related to customer services and trouble call management were supported.

3.3.7 IEC 61968-9 Application Integration at Electric Utilities - System Interfaces for Distribution Management - Part 9: Interface for Meter Reading and Control

Part 9 [57] of the standard series focuses interfaces for meter reading and control. Typical use cases are:

- meter reading
- meter control
- meter events
- customer data synchronization
- customer switching

IEC 61968-9 defines how metering systems have to be integrated with other systems and business processes within the application area of IEC 61968 standard

[2] As presented in May 2011 at the WG 14 meeting in Prague.

series. Thereby, a metering system consists of Automated Meter Reading (AMR)- systems and traditional, manual systems. This standard should be used for the integration of metering systems and other systems within utility companies. However, communication protocols are not in the scope of the standard. The resulting main advantage is, that the standard is not influenced by the development of the future metering infrastructure and thus there is almost no risk that the standards becomes obsolete.

3.4 Component Interface Specifications

In [26] the structure of the part 400 documents of the IEC 61970 standard series is explained as follows:

1. **Part 401:** CIS Framework - This document provides an overview of the Part 4 CIS series of standards and an explanation of how to use the standards in system implementations and system integration projects.
2. **Parts 402-449:** These parts are reserved to specify the application independent services to be supported by component interfaces. These specifications describe in narrative terms with text, UML notation, and Interface Definition Language (IDL) the API that is standardized. These specifications define the generic services that can be used by any application to exchange information with another application or for public data access.
3. **Part 451[3]:** CIS Information Exchange Model Specification Guide - This document provides an overview of the process used to define and document information content and exchange mechanisms as well as examples of system integration using the CIS standards.
4. **Parts 452-499:** These parts are reserved for specifications that address the specific application category dependent information exchange requirements as encapsulated in an Information Exchange Model (IEM). These IEM specifications define the information content of the standard information exchanges between applications as well as what Part 402 - 449 Service is used to transfer data during the course of component interactions. This data is defined as messages but may be exchanged in a variety of ways, including being published/subscribed as events or responses following a request. For every component dialog defined in an IEM, the context of the messages exchanged and services used are identified. Supporting documentation includes use cases and event sequence diagrams.
5. **Parts 5xx:** A separate series of documents (i.e., Part 5xx) describe mappings to specific technologies for the implementation of these standards. To ensure interoperability, there must be a standard mapping for each interface to each technology. For example, if Web Services is the chosen implementation technology, then there needs to be a standard mapping of the publish/subscribe services specified in the CIS document to XML compliant with Web Service standards.

In this section, the standards introduced in the first two bullet points are presented.

[3] This is now part 450 (see section 3.5.1) and part 451 has a new focus (see section 3.5.2).

3.4.1 *IEC 61970-401 Energy Management System Application Program Interface (EMS-API) - Part 401: Component Interface Specification (CIS) Framework*

A framework for the Component Interface Specifications (CIS) for the EMS-API is defined in part 401 [35]. Thereby, CIS define interfaces which are implemented by those components or applications wanting to exchange data with other components and applications in a standardized way or to gain access to publicly available data. The specifications are allocated to the following sub parts of the standard series. Parts 402 to 449 specify generic services for data exchange and parts 451 to 499 define information content. Figure 3.6 shows the specifications and other related standards from OMG and OPC.

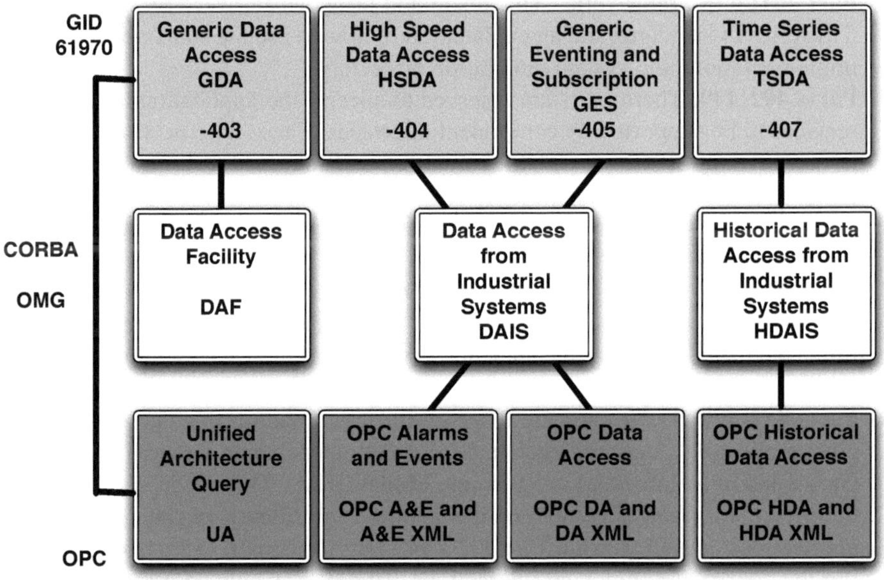

Fig. 3.6 Overview on the Component Interface Specifications and related standards [35]

3.4.2 *IEC 61970-402 Energy Management System Application Program Interface (EMS-API) - Part 402: Common Services*

Part 402 [51] defines common services, which are the basis for the GID. These API-services are required as basic services for the following parts of the standard series. This implies that all services defined in part 402 must be implemented if the following interface should be realized:

- Generic Data Access (GDA)
- High-Speed Data Access (HSDA)
- Generic Eventing and Subscription (GES)
- Time Series Data Access (TSDA)

The semantics used for the interfaces is described in IEC 61970-301 and IEC 61968-11. The common services can be divided into the three following categories [51]:

- Resource identifier: Generic services to identify classes, class attributes and object instances in systems.
- Resource descriptions: Generic services to encode values associated with classes, class attributes and object instances.
- Views: Services to represent classes, class attributes and object instances by hierarchies and tree structures.

Although the standard series is focusing on control centers the specified services could also be implemented for other systems like DMS, asset management systems, GIS and Object Management System(OMS). They are used to couple these systems via CIM-conform interfaces.

3.4.3 IEC 61970-403 Energy Management System Application Program Interface (EMS-API) - Part 403: Generic Data Access

This part [52] covers generic data access using the common services specified in part 402. GDA defines services which can be used to access CIM-conform entities. Thereby, GDA permits clients to use data provided by other components like databases and other systems without having knowledge about their internal logical data storage schema. The services are designed to realize a synchronous and not real time based communication of complex data structures among components. A typical use case for GDA interfaces is the exchange of mass date between a database and an analysis application identifying the state of the grid based on the current data from the database, storing the result in the database again and notifying the initiating system.

Due to reducing the complexity and utilizing database independent technologies, the services are defined as generic as possible. Their capability is restricted to read, update and notify changes of objects. Thus, GDA could be used as Enterprise Information Integration Adapter (EIIA).

3.4.4 IEC 61970-404 Energy Management System Application Program Interface (EMS-API) - Part 404: High Speed Data Access (HSDA)

Part 404 [44] deals with a common interface for High Speed Data Access (HSDA). Latency's of Local Area Networks (LAN) are considered, so that they can be used for efficient data exchanges. This part is derived from the DAIS DA (Data Acquisition from Industrial Systems Section Data Access) specification which again is based on OMG Data Access Facility (DAF) and OPC DA. DAIS DA is a platform-specific model related to Common Object Request Broker Architecture (CORBA).

OPC DA is also a platform-specific model but related to Microsoft Component Object Model (COM). Part 404, however, describes a platform independent model as a complement to a platform specific model. The functionalities of both platform specific models are described independently, consequently, they can be used as basis to develop further platform specific models or as introduction to already existing platform specific models.

The HSDA interface is intended to be interoperable with the other interfaces specified in the IEC 61970 standard series. Hence, it is possible to request the same information like object identifier, attribute names and identifier, and class names and identifier via the different interfaces. Implementations of specific platform specific models are covered in other parts of the series. Specific mappings are and will be described within the parts 5XX.

3.4.5 IEC 61970-405 Energy Management System Application Program Interface (EMS-API) - Part 405: Generic Eventing and Subscription (GES)

Within part 405 [45] another interface is described. It also considers the latencies of LANs and enables efficient data exchange. The interfaces shall be used to exchange event messages and the API provides the primary tool for application integration. Compared to other APIs, this one does not cover high performance, real-time interactive requirements of applications in running systems or request/reply oriented generic data access.

Analogous to part 404, this part also describes a platform independent model which provides functionalities of different platform specific models. In this case the addressed platform specific models are Data Acquisition from Industrial Systems Alarms and Events (DAIS A&E), OMG DAF and OPC A&E. DAIS A&E is a CORBA based model and OPC A&E is based on Microsoft COM. As intended for part 404, this part should also be interoperable with the other IEC 61970 interfaces.

Beside the aforementioned platform independent model which es related to SCADA A&E, part 405 includes another important section addressing GES in a more general, SCADA-independent way.

3.4.6 IEC 61970-407 Energy Management System Application Program Interface (EMS-API) - Part 407: Time Series Data Access (TSDA)

IEC 61970-407 [46] deals with an interface for time series data access. Regarding the interoperability with the other IEC 61970 interfaces, the reference to the parts 5XX and the reference model as well as the efficient data exchange via LANs there is no difference between this part and the parts 404 and 405. The difference are the platform specific models whose functionalities are covered by the described platform independent model. Namely, the functionalities of HDAIS (Historical Data

Access from Industrial Systems), OMG DAF and OPC Historical Data Access (HDA) are emulated.

3.5 Information Exchange

Within this section the CIM-based information exchange is introduced as it is specified in the parts 45x of the standard series IEC 61970. The parts described address specific application category dependent information exchange requirements. These requirements are part of an IEM. In short, the IEM specific actions define the content of the standard information exchange between different applications or services mentioned in chapter 3.4. Most of data is defined as messages but the possibilities of exchanging can be different, for example publish/subscribe events can be used. Each identified component in the IEM will be defined including the context of the exchanged message as well as the services used in the process. The included documentation covers use cases and supporting sequence diagrams.

3.5.1 IEC 61970-450 Energy Management System Application Program Interface (EMS-API) - Part 450: CIS Information Exchange Model Specification Guide

Since the parts 450 - 499 deal with the information content of messages and also define which services are applied by components or systems in order to exchange information with other components, IEC 61970-450 [26] gives an overview on as well as guidance for these standards. Formerly, the content of this standard was covered by part 451 (see section 3.5.2) until it gets a new focus. Within this standard requirements are identified and established which are related to those standard interface specifications being based on the application integration architecture (see section 3.1.3.

3.5.2 IEC 61970-451 Energy Management System Application Program Interface (EMS-API) - Part 451: CIS Information Exchange Model For SCADA

Currently, IEC 61970-451 [29] is still a working draft as its original content was shifted to part 450. Now, the focus is on an information exchange model for SCADA systems whereas also here, only the formats for message exchange are defined and not the according transports and interfaces. It is important to notice that both transport and interface discussions are not normative in this context but in later versions of the specifications the sequences of the messages will become normative.

3.5.3 *IEC 61970-452: Energy Management System Application Program Interface (EMS-API) - Part 452: CIM Transmission Network Model Exchange Profile*

The standard IEC 61970-452 [58] defines CIM-profiles which are used to exchange static power system data[4]. The models are exchanged between entities like utilities and security coordinators being involved in an interconnected power system. The goal is that all parties will have access to those data of their neighbor systems that are needed for state estimation and power flow applications. Therefore, the information from part 452 defining the RDF based CIM XML model exchange format, should be used to exchange power system model data. A XML file used for model exchange is only than valid if it is compliant to those conventions which are established within the standard, whenever it is necessary.

The standard initially relies on the originally data requirements identified by the North American Electric Reliability Council (NERC) CPSM Data Exchange Working Group (DEWG). Even those requirements are based on industry practices, it is necessary to extend them to a certain degree in order to facilitate model exchanges including parameters common to breaker-oriented applications.

The standard can be read from the two following perspectives:

- **Data producer:** From export software's point of view the standard defines the minimum set of CIM objects (CIM classes, attributes and associations) which must be included in a XML file used to exchange power system models. Moreover, the standard does not define how the network has to be modeled. As an example, if the modeled network does not include any breakers, the exported XML file should not contain instances of the CIM class Breaker. Furthermore, if the modeled network includes breaker, the exported XML file should contain instances of the CIM class Breaker, at least including the required related attributes and associations.
- **Data recipient:** Importing software must be able to interpret those CIM objects which are defined in the profile in order to import the power system model. As the profile defines a minimum set of CIM objects, the models represented by the XML files could contain any further CIM objects as long as the resulting data is compliant to the CIM RDF Schema and the conventions established in this standard. Besides, the required CIM objects the most likely are labeled as recommended or conditionally required.

3.5.4 *IEC 61970-453 Energy Management System Application Program Interface (EMS-API) - Part 453: CIM Based Graphics Exchange*

Sub-part 453 [47] of the standard series deals with CIM-based graphics exchange. As all sub parts from 450 to 499 also part 453 addresses the exchange of data and

[4] Currently only the Equipment Profile has been defined.

exchange mechanisms between control system components on an abstract level. The sub standards comprises use cases for the exchange of schematic graphics like single-phase representations (single-line diagram) in control systems or GIS. Moreover, schemas and mechanisms are defined, describing how these graphics can be aligned to CIM-based data from IEC 61970-301 and IEC 61968-11. The generic use cases include coupling of control systems among each other, to GIS-systems as well as planning and trading systems. In addition to the meta model covering the annotation of graphical data to the CIM model, the serialization of the models using the XML format (as Scalable Vector Graphic (SVG)) is defined.

3.5.5 IEC 61970-454 Energy Management System Application Program Interface (EMS-API) - Part 454: Naming Service Specification

IEC 61970-454 [37] contains a naming service specification and thus defines an information model and related services used to access a business object naming registry. In this context, business objects are understood as instances of CIM-classes. One main objective is to handle the use of multiple names for the same resource (named instance of a specific class or type) as well as the need for cross-references between those names. The design of the cross references is enabled by a Resource Name Repository (RNR) which also provides an example addressing the issue by the use of the IEC 61970 interfaces. The following three objectives are pursued by the creation of resource name cross-references [37]:

- Support the merging of overlapping models, e.g., merge an imported CIM XML file with an existing model.
- Provide a cross-reference between resource names from different data models.
- Help to assure consistency of data models as changes are made.

Even if the primary need for cross-references is for exchanging power system models, the approach described in this standard can also be used to resolve names for other resources like assets or work orders for both cases, between and within utilities. Furthermore, the approach is able to support many business processes because it is based on vendor neutral and open standards.

3.5.6 IEC 61970-455 Energy Management System Application Program Interface (EMS-API) - Part 455: Model Population Interfaces

The standard IEC 61970-455 is planned to cover model population interfaces but its development is currently on hold until an appropriate working draft is provided[5].

[5] As presented in February 2011 at the WG 13 meeting in Santa Clara.

3.5.7 IEC 61970-456 Energy Management System Application Program Interface (EMS-API) - Part 456: Solved Power System State Profiles

IEC 61970-456 [64] supports power flow, state estimation and similar applications by defining CIM-profiles being required for descriptions of steady-state solutions of power system cases. Thereby, the proposed solution refers to power system models being conform to part 452 and they are facilitated by included, informative use cases. The described profiles are each related to a specific component (topology derived from switch positions, measurement input in the case of state estimation and the solution itself).

Similar to part 452 also this standard can be read from two perspectives:

- **Data producer:** In this case, the standard describes how a model export software can describe an instance of a network case with the intention to make it accessible for other programs.
- **Data recipient:** Here, the standard describes what capabilities in terms of interpretation a model importing software must have in order to consume solution cases.

There are many situations in which this standard could be applied and thus many use cases exist for which the utilization of the standard differs. So implementers should consider the use cases they want to cover. Depending on the chosen use cases, they have to cover different options. For example, if the standard is used to exchange starting conditions (instead of solved conditions), consumer applications must be able to cope with, e.g., unsolved states.

3.5.8 IEC 62325-450 Deregulated Energy Market Communication - Part 450: Profile and Context Modeling Rules

The IEC 62325-450 [65] standard defines design rules to create XML implementation schema to use for interfaces between different market actors. It describes how the UML model objects have to be interpreted and transformed into a schema and thus allows for an efficient integration of information interchanges between electricity market participators. It ensures modeling consistency and avoids ambiguity between objects by providing a clear understanding on what they are based on within the CIM.

Figure 3.7 illustrates how this part can be positioned in the IEC 62325 modeling framework.

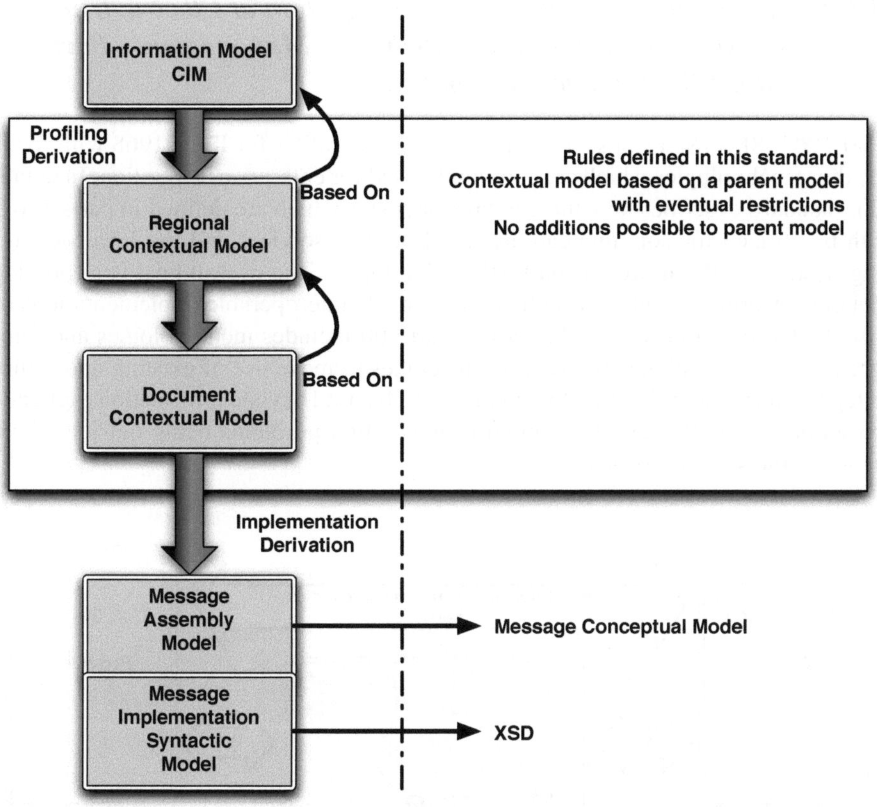

Fig. 3.7 General overview on the focus within the modeling framework principles [65]

3.6 Technology Specifications

Since one paradigm for the development of the CIM is the independence from specific platforms and technologies, it is necessary to specify certain mappings to make the CIM applicable. In this section mappings are introduced which are used in the context of different use cases. They address the use of CIM-based information exchange for ESB implementations, harmonized with MultiSpeak, serialized with RDF to exchange topologies, mapped to OPC UA for realizing web service and binary data exchanges, serialized as OWL ontology, and in combination with Electronic Business using XML (ebXML) for market communications.

3.6.1 IEC 61968-100 Application Integration at Electric Utilities - System Interfaces for Distribution Management - Part 100: ESB Implementation Profile

Part 100 [68] considers a set of implementations profiles for IEC 61968 with focus on those technologies which are commonly used for enterprise integration. In detail, this standard describes how the message payloads which are defined in parts 3 to 9 can be utilized for communications based on web services and the Java Messaging System (JMS) as well as on ESB technologies. The overall goal is to provide detailed information which is sufficient to enable interoperable implementations of IEC 61968-based messages. Moreover, part 100 includes methodologies and integration patterns which could serve as leverages to make use of existing and future integration technologies. So, this standard can be used beyond information exchange based on IEC 61968 and be applied to integration problems being outside of the scope of the standard series.

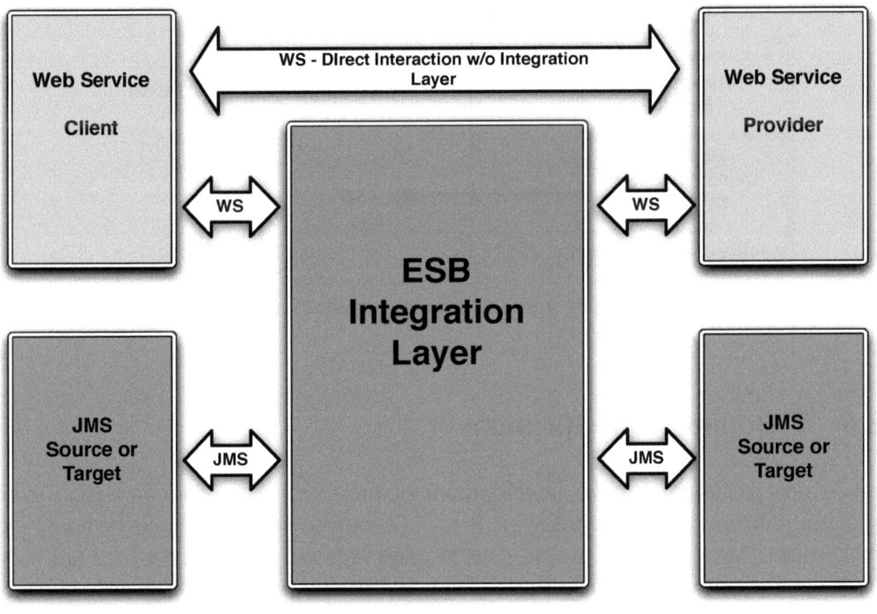

Fig. 3.8 Scope of IEC 61968-100 [68]

Figure 3.8 depicts where IEC 61968-based messages can be used in the context of web services and JMS. By integrating an ESB it is possible for the initiator of an information exchange to use web services whereas the receiver uses JMS and vice versa. The ESB as an integration layer also enables on to many information exchanges based on publish/subscribe integration patterns as well as key functionalities like delivery guarantees.

The scope of IEC 61968-100 can be summarized by the following points [68]:

- integration patterns which support IEC 61968 information exchanges
- design of interfaces for using web services
- design of interfaces using JMS
- definition of standard design artifacts and related templates

Finally, it is important to notice that the described implementation profile is not the only possible one for IEC 61968 information exchange, so other implementation profiles can be defined. Furthermore, prescribing implementation details as required for security is outside the scope of this standard.

3.6.2 IEC 61968-14 Application Integration at Electric Utilities - System Interfaces for Distribution Management - Part 14: MultiSpeak - CIM Harmonization

Part 14 [69] aims at the harmonization of CIM and MultiSpeak (V 4.1) which is also a leading standard for software interoperability in the energy domain like the CIM. However, it is mainly used in the United States among municipal and cooperative utilities and vendors which serve these utilities. Furthermore, it is maintained by the MultiSpeak Initiative. The goal of the harmonization is to express the capability currently inherent in MultiSpeak as a CIM profile. Mapping on profile level means that each CIM profile has a mapping to a corresponding MultiSpeak element. This enables both, the opportunity to translate interfaces based on one of the two standards to the other and revealing alignments as well as identifying gaps. The exercise also comprises a practical example using an Enterprise Service Bus (OpenESB) to show the interoperability of interfaces reached by the mapping. A XML mapping tool is used to create the mapping and also the testing of an import of CIM-based and MultiSpeak artifacts into an OWL ontology for analysis purposes is discussed.

Several sub parts of part 14 are planned to introduce different profiles. Part 14-1 will cover the general harmonization of CIM and MultiSpeak, whereas, e.g., part 14-1-9 is a mapping from part 9 dealing with profiles for metering messages to MultiSpeak message payloads, and part 14-1-13 is a mapping to part 13 covering the Common Distribution Power System Model (CDPSM) profile.

3.6.3 IEC 61970-501 Energy Management System Application Program Interface (EMS-API) - Part 501: Common Information Model Resource Description Framework (CIM RDF) Schema

IEC 61970-501 [38] specifies a specific technology mapping. It maps the conceptual UML data model (defined in IEC 61970-301 and IEC 61968-11) and its schemes to a machine-readable XML representation. Therefore, the RDF specification is used. The format and the rules were set, which enable the generation of a machine-readable CIM representation. This mechanism should be used by vendor independent applications to gain access to CIM meta data. This access is based on a general format and the standardized services defined in the IRM or CIS. Beside this primary objective two secondary objectives were pursued. On the one hand, the capability of CIM revision control should be supported and on the other hand, the mechanism should be easily expandable to support location dependent services. This means services, which must have knowledge about the real location of a physical object, so an abstracted model is not sufficient for them. The aspired solution has the following characteristics [38]:

- is machine-readable and human-readable, although primarily intended for programmatic access
- can be accessed using any tool that supports the Document Object Model (DOM) application program interface
- self-describing
- taking advantage of the current W3C Web standards

3.6.4 IEC 61970-502-8 Energy Management System Application Program Interface (EMS-API) - Part 502-8: Web Services Profile for 61970-4 Abstract Services

This sub part of the standard series has the working title "CIM Data Services" [53], and is in CD stadium. It mainly deals with a mapping between CIM objects and the OPC Unified Architecture information model to enable CIM-based Web Service communication within SOA. The OPC UA describes a platform-independent SOA and it is maintained by the OPC Foundation as well as adapted by the IEC as series 62541. Beside the Web Service communication the CIM-UA-mapping also allows a binary based communication for time critical purposes. Within the part 502-8 mappings are introduced for CIM classes, attributes and associations, so it is possible to use all given CIM semantics for an UA based server-client architecture.

3.6.5 IEC 61970-505 Energy Management System Application Program Interface (EMS-API) - Part 505: OWL Schema

IEC 61970-505 is a planned standard describing an OWL schema. However, for this part a working draft has to be prepared and a NWIP is still needed[6].

[6] As presented in February 2011 at the WG 13 meeting in Santa Clara.

3.6.6 *IEC 61970-552 Energy Management System Application Program Interface (EMS-API) - Part 552: CIM XML Model Exchange Format*

The standard IEC 61970-552 [59] specifies the CIM/XML model exchange format. Since the data model defined in IEC 61970-301 and IEC 61968-11 is represented using UML - usually utilized to specify, visualize and document systems in an object-oriented way - to analyze and design the model, it has to be transformed to make it applicable for software programs. The transformations target must be a schema which allows to use a programmable interface. This is done in IEC 61970-501 by translating the UML format into in XML representation of a RDF Schema specification. Now, part 552 defines how this CIM RDF schema could be used to exchange power system models, mainly driven by the need of exchanging transmission network modeling information for power system security coordination. Thereby, XML is used as defined in the parts 45x focusing on profiles, e.g., IEC 61970-452 describing the CIM Transmission Network Model Exchange Profile, and defining the use cases and requirements for this standard. Therefore, the information from part 501 are used as a meta model framework to construct XML documents including power system modeling information.

The supported mechanism addresses software from different vendors covering the production and consumption of CIM-based modeling information. Hence, the solution has the same characteristics as described in section 3.6.3.

3.6.7 *IEC 62325-501 Framework for Energy Market Communications - Part 501: General Guidelines for Use of ebXML*

IEC 62325-501 [32] describes general guidelines how to integrate the ebXML technologies and architecture. Also, different migration scenarios as well as implementation examples will be provided, too. The recommend profiles for this scenarios can be found at IEC 62325-502 at chapter 3.7.4. This part does not define a standard itself but references existing standards from the Internet Engineering Task Force (IETF).

3.7 Profiling and Testing

Profiles are an established means to cope with the large size of the CIM data model. The general background and use of different profiles is explained in detail in section 3.8. In this section one important profile is introduced because it is part of the IEC 61968 standard series. The profile is called CDPSM and specifies the set of CIM objects that is needed to exchange distribution grid models. Furthermore, profiles are introduced which are part of the IEC 62325 standard series dealing with market communications. The second part of this section covers interoperability testing.

However, the according standard is in a very early stage and does not provide many information so far.

3.7.1 IEC 61968-12 Application Integration at Electric Utilities - System Interfaces for Distribution Management - Part 12: Compliance and Interoperability Testing

IEC 61968-12 formerly contained use cases for the services used for CIM-based message exchange and described in the parts 3 - 9. Now, the use cases will be directly included in the related parts. Thus, part 12 is a new working draft considering compliance and interoperability testing[7].

3.7.2 IEC 61968-13 Application Integration at Electric Utilities - System Interfaces for Distribution Management - Part 13: CIM RDF Model Exchange Format for Distribution

One of the CIM-profiles discussed in section 3.8 is standardized in IEC 61968-13 [49]. The standard includes a CIM RDF model exchange format for distribution defining the CDPSM. In a nutshell, profiles represent a defined set of classes and their attributes as well as associations connecting the classes, defining a subset of CIM-objects needed for a specific use case (see section 3.8.1). This facilitates the definition of interfaces for developer and system manufacturers. Due to the restricted and application-specific size, interoperability problems will arise less often, because the classes used for a specific case are exactly defined and thus almost none alternative definitions are left. Part 13 specifies those classes and relations which are needed to serialize distribution grids in an European context (UCTE). For serialization, XML and RDF formats are used. This guarantees the interoperability of exchanged topology data among different systems. So called CIM-Interoperability-Tests are performed and attended by different vendors to proof the interoperability of their products in practice. For this purpose, the CDPSM is defined as RDF-schema (technically similar to XML-schemas used to verify XML-messages) allowing an automated verification of the serialization of the data instances.

3.7.3 IEC 62325-351 Framework for Energy Market Communications - Part 351: CIM European Market Model Exchange Profile

The CIM European Market Model Exchange Profile defined in Part 351 [74] of the IEC 62325 Standard defines a subset of the CIM objects as a Regional Contextual Model to cover the market information interchange requirements for the European region. This profile has been generated using the rules defined at the IEC 62325-450

[7] As presented in May 2011 at the WG 14 meeting in Prague.

part described at chapter 3.5.8 and is based on the original CIM artifacts. However, a particular artifact may be refined respecting a set of defined rules. The specific regional artifacts themselves cannot contradict the CIM artifacts on which they are built. The result of this standard is a specification of a package which provides a logical view of functional aspects of the European Electricity Market Management.

3.7.4 IEC 62325-502 Framework for Energy Market Communications - Part 502: Profile of ebXML

The 502 Part [33] of the IEC 62325 Standard defines a profile within the market framework based on ebXML. It does not define a standard but only references existing ones from organizations like W3C, OMG and UN/CEFACT.

3.8 CIM Profiles

3.8.1 What Are CIM Profiles?

As often only parts of the CIM are used within a project or company, the concept of CIM Profiles was introduced. CIM Profiles are subsets of the CIM. They include only essential classes and associations which are needed for a specific scenario or task. In addition, further restrictions like cardinalities of associations are specified. In summary, CIM Profiles limit the scope of the CIM by allowing only to use selected elements of the CIM (like classes) and adding further restrictions. Through CIM Profiles the use of the CIM becomes more easy. At the moment several discussions are undergoing regarding the use of CIM Profiles. Industry is asking for conformity and compliance test of CIM based data exchange. Because of the nature of the CIM offering a flexible and generic data model where all attributes and associations are in principal optional, a conformity check against CIM is not convenient because it will always return true as long as CIM classes and attributes are used. CIM Profiles restrict the use of the CIM and distinguish between mandatory and optional attributes and cardinalities of associations for given use cases. Therefore, interoperability (IOP) tests and conformity tests should be performed against profiles. Standardization of the CIM is at the moment aiming at keeping the CIM as flexible and generic as possible and restricting its use by defining CIM profiles for given use cases to make it testable.

On the one hand some companies use intra-corporate profiles, and on the other hand large profiles exist which are partly standardized and widespread within the utility domain. In the following the most common CIM Profiles are listed:

- **CPSM:** The Common Power System Model (CPSM) is used in the USA for the exchange of transmission system models [58].
- **CDPSM:** The Common Distribution Power System Model (CDPSM) is used in Europe for the exchange of distribution power system models [49].

- **ENTSO-E:** The ENTSO-E[8] profile is used in Europe for the exchange of transmission system models.
- **ERCOT:** The Electric Reliability Council of Texas[9] (ERCOT) profile is an intra-corporate data model.

3.8.2 *Creating Your Own CIM Profile*

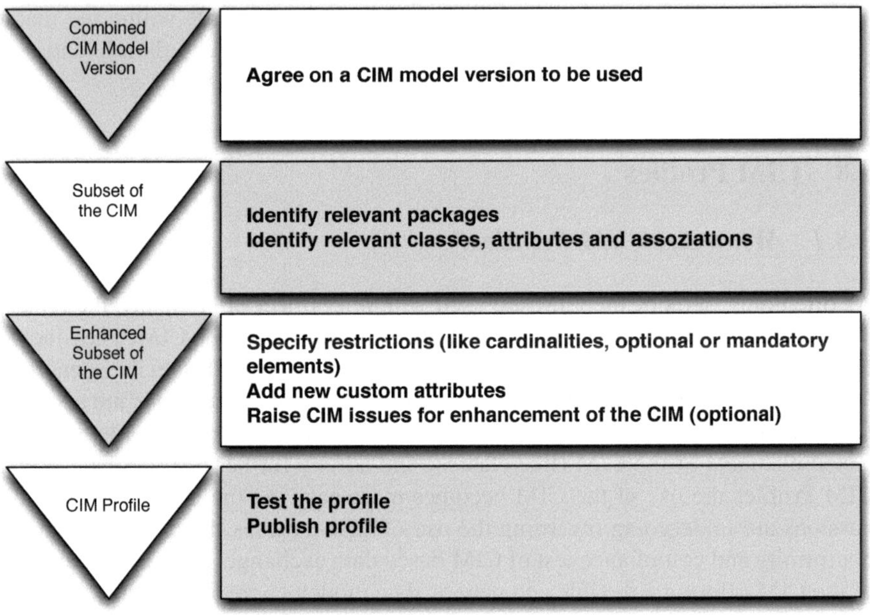

Fig. 3.9 CIM Profile development

Figure 3.9 illustrates the basic steps to undertake within CIM Profile development. First a CIM model version to be used for profile definition has to be agreed on. Based on this relevant packages, classes, attributes and associations have to be identified to build a relevant subset of the CIM for a specific use case. Then restrictions like cardinalities for associations as well as optional and mandatory attributes need to be specified. Within this usually enhancements of the CIM are needed or problems occur so that submission of CIM issues might be necessary. After definition of an own CIM profile is completed test with applications should be conducted and the CIM profile provided. In case a public CIM profile is defined to become an international standard, IOP-Tests with different vendors should be performed.

[8] http://www.entso-e.org/
[9] http://www.ercot.com/

3.9 Data Model Extensions

As mentioned in this chapter, the CIM includes a comprehensive data model, but sometimes these model is insufficient to meet certain requirements. The origin of this shortage is that it is impossible to include all imaginable objects. The model would be to big and many objects cannot be foreseen. Because of this a extension of the data model is sometimes necessary. Unfortunately, there is no ideal way to extend the data model. In this chapter we will show one possibility, which is recommended within the CIM user group and can be done with the Sparx Enterprise Architect UML editor (further described in chapter 6.3.1, which is used to model and maintain the CIM data model.

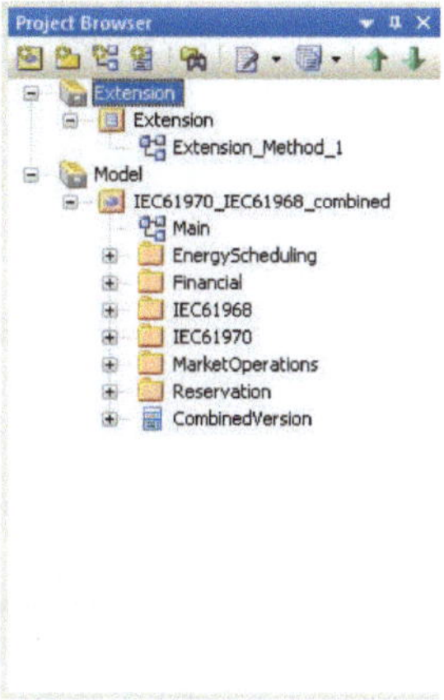

Fig. 3.10 Enterprise Architect Project Browser after adding a new extension model

Before extending the CIM, a thorough search within the data model is necessary and a communication with other organizations using CIM can be of help. If no solution to model the required objects or characteristics can be found, an extension is inevitable but should be most non-intrusive to the original data model as possible. To ensure this, it is necessary to abstain from modifying the original CIM data model but to extend it by whole classes or by using inheritance to add required attributes to a existing class. Figure 3.11 shows a possible workflow to create the desired extension.

Basically, there are two possibilities to extend the CIM. Either there is a class to use at a base and extend only attributes or to create an entire new class. To minimize the semantic changes on the original CIM model, both options require the preparation of an additional model.

To create a new model, but use the CIM base model simultaneously, the original data model can be loaded into Sparx Enterprise Architect and add a new model in the project. Figure 3.10 shows a snippet of the project browser after adding a new extension model and the corresponding view and diagram.

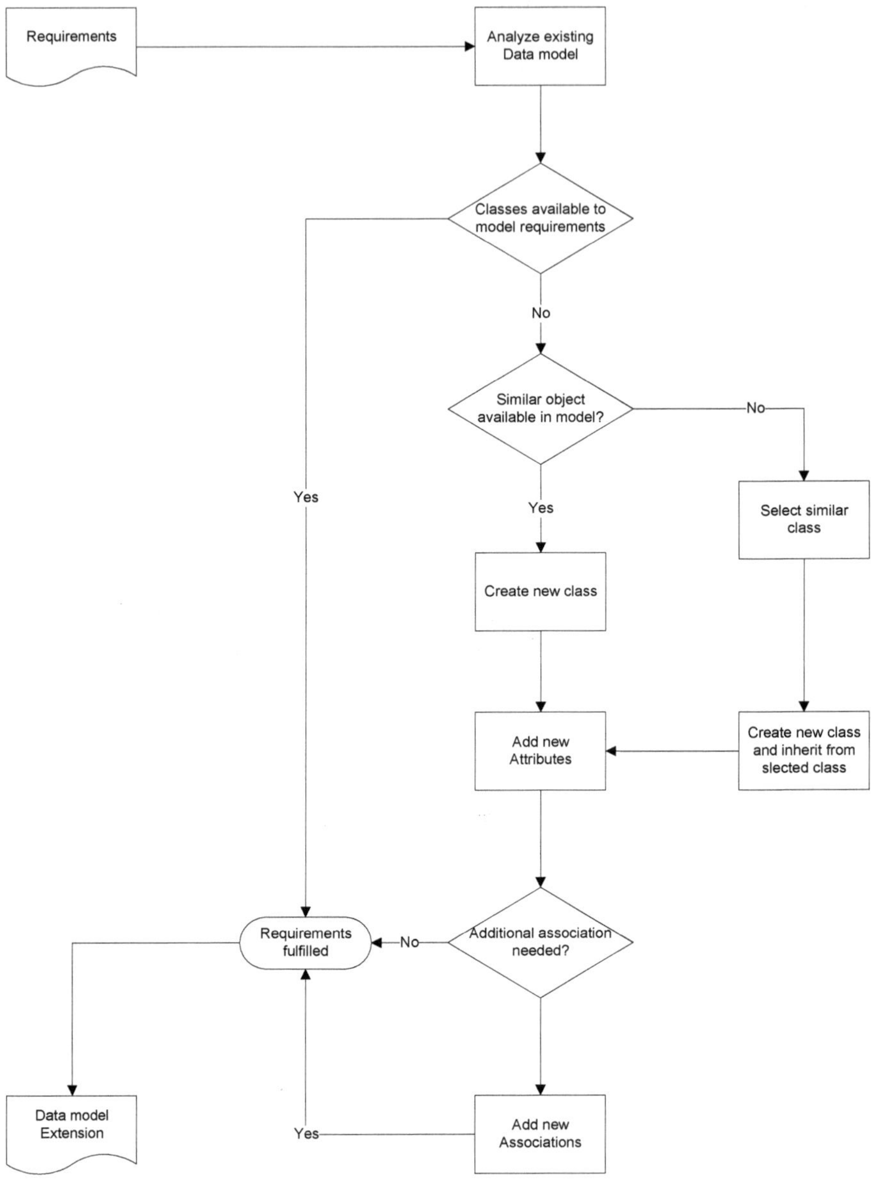

Fig. 3.11 Workflow to extend the CIM [110]

To extend the CIM by inheritance (only possible if there is a suitable base class), it is necessary to drag-and-drop the base close into the new diagram. In there a new class, with a semantically appropriate name should be created and an inheritance association should be created between those classes. It is now possible to extend the new class with attributes. Figure 3.12 shows the result of extending the CIM data model by inheritance. The class *SwitchExt* was created and inherits from the CIM class *Switch*. Additionally, the attribute "'x"' has been added to the class *SwitchExt*.

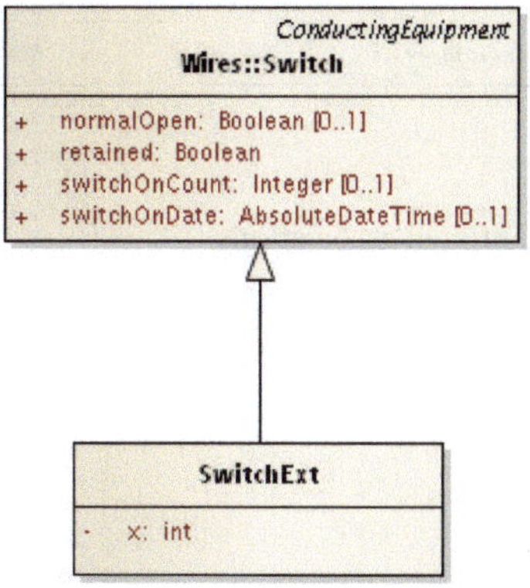

Fig. 3.12 Extending the CIM by using inheritance

In some cases the creation of new associations is needed. In the example case in figure 3.13 the previous extension has been enriched with an additional association including multiplicities, whereas the class *AnotherClass* stands for every possible class out of the CIM data model or another extended class. The association ends should also be named.

To extend the data model without a semantically suited class the previous steps are almost the same. The only difference is to drop the inheritance. All other steps remain identical.

This created extensions can now be exported to the XML Metadata Interchange (XMI) format available in Enterprise Architect and can therefore be imported into different tools. Some tools like CIMTool (chapter 6.2.3) and CIMBench (chapter 6.2.1) are supporting XMI extensions of the CIM explicitly. These extensions can also be shared with other parties, to ensure a higher interoperability. Some extensions have also found the way into the standardization process to standardize them for a wider user group and to integrate them into the CIM data model.

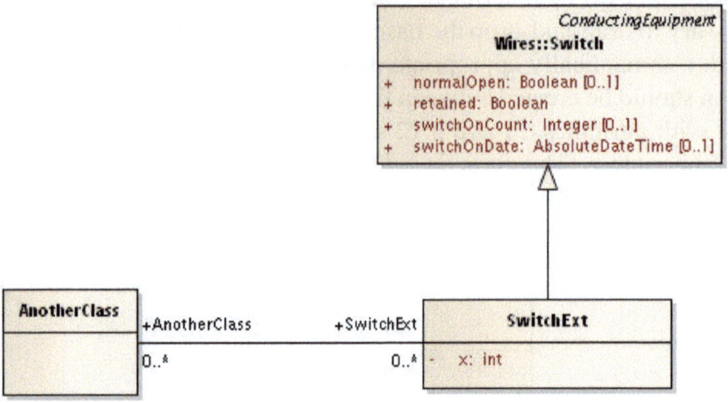

Fig. 3.13 Extending with additional associations

Chapter 4
Utilization of the CIM

4.1 XML Based Message Exchange

Despite the fact that CIM messages can be used on different technologies, this chapter focuses on the XML based message transportation. The IEC 61968 "System Interfaces for Distribution Management" standard series is intended to facilitate inter-application integration of various distributed software application systems. Loosely coupled applications with a high rate of heterogeneity in languages, operating systems and protocols and a need to exchange data are the target of this standard series [41] as graphically clarified in figure 4.1. This is the main use case for using XML based messaging with CIM. Therefore, the IEC 61968 and the subparts will be dominant in the following parts.

4.1.1 General Introduction in CIM Based Messaging

Communication between application requires compatibility on the two levels [41]:

- Message formats and protocols
- Message contents must be mutually understood, including application-level issues of message layout and semantics

The IEC 61968-1 standard defines among other things abstract middleware services to support the communication between applications. These abstract services can be deployed, by mapping them to commonly available services from various messaging technologies like Web Services (WS) (see chapter 2.4) and middleware like Enterprise Services Buses (ESB) (see chapter 2.4). The IEC 61968 standard series even provides a specification of an implementation profile for the application of the other parts of 61968 using common integration technologies, including JMS and web services in the IEC 61968-1-1 "Implementation Profile for IEC 61968" standard (see [67]).

Because of the practical usability of the IEC 61968-1-1 standard, the following part will focus on to give the reader a further understanding of interface and

M. Uslar et al.: The Common Information Model CIM, POWSYS, pp. 107–124.
springerlink.com

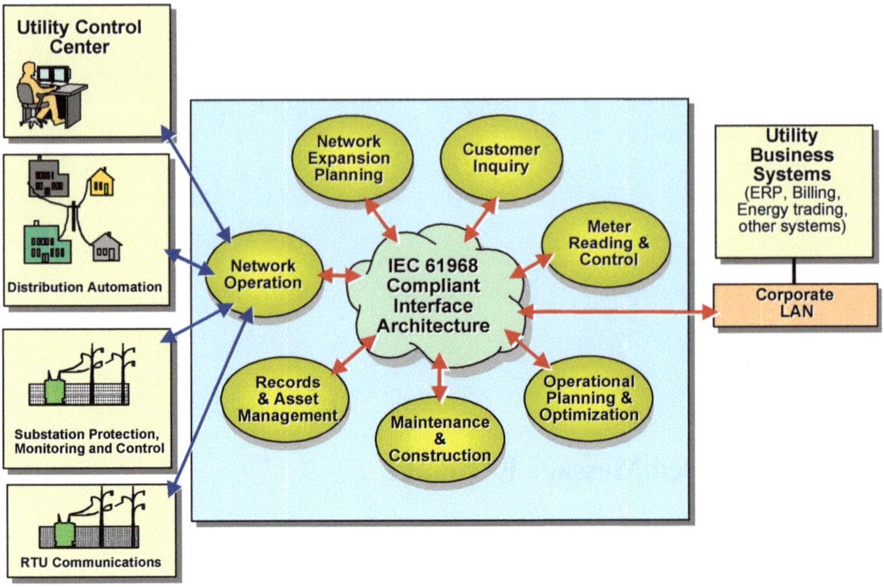

Fig. 4.1 IEC 61968 compliant interface architecture [41]

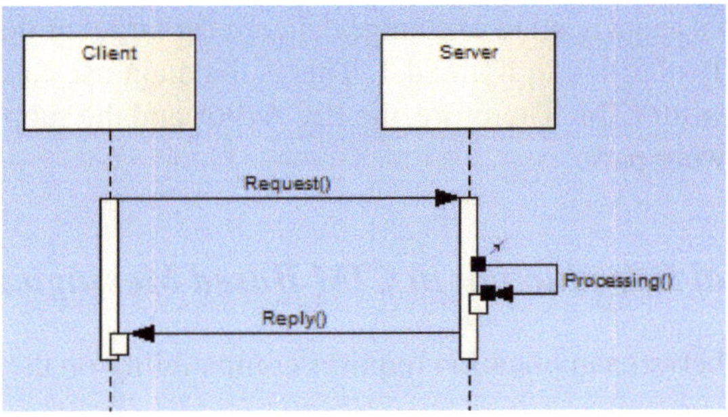

Fig. 4.2 Simple request/reply message use case [67]

message implementation in CIM. Therefore, most parts will be concerned with ESB technologies, respectively underlying technologies like Web Services.

The use cases for messages cover many possible interactions between different components within a set of systems to support business processes. These use cases range from simple request/reply use cases (figure 4.2), where a client send a request message directly and the server send a reply message after processing, up to complex messages with potentially many asynchronous replies (figure 4.4). These complex messages can consist of a chain of messages to different systems connected to an ESB. Even event messages (see figure 4.3 can be part of complex messages, which are another form of messages which only reaches the systems subscribed to an ESB.

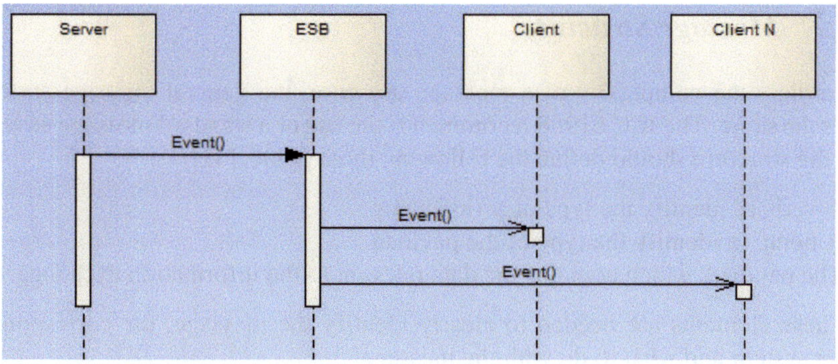

Fig. 4.3 Event message use case [67]

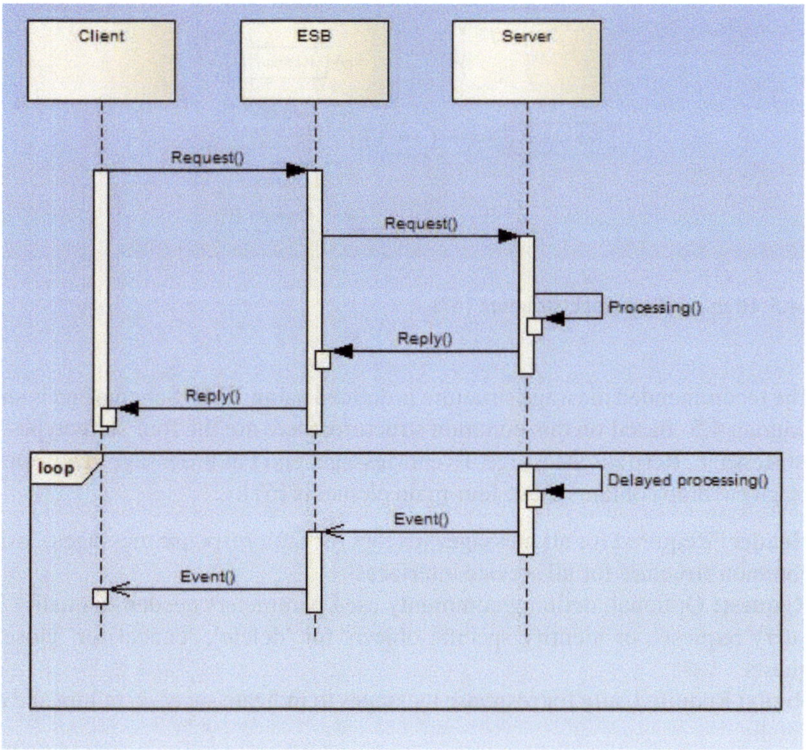

Fig. 4.4 Complex message use case [67]

4.1.2 Message Structure

Regardless the complexity of a message use case, the general message structure stays the same. The IEC 61968 recommends the use of a standard message envelope and the structure should define the following three points [41]:

- A verb, to identify the type of action being
- A noun, to identify the type of the payload
- The payload, which contains the data relevant to the information exchange

These elements are needed to clearly identify the message, the corresponding target system and what to do with the message.

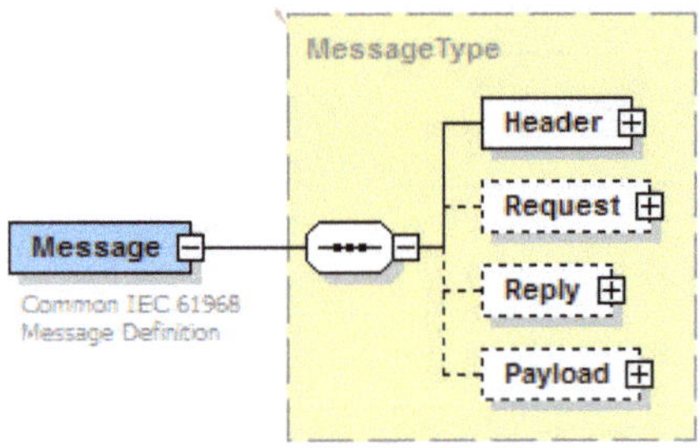

Fig. 4.5 High-level message structure [67]

The recommended message structure is defined using XML Schema and is shown in diagram 4.5. Based on this common structure, there are the four stereotypes: RequestMessage, ResponseMessage, EventMessage, and FaultMessage. The common message envelop contains these four main elements [67]:

- **Header:** Required for all messages (except for fault response messages), using a common structure for all service interfaces
- **Request:** Optional, defining commonly used parameters needed to qualify 'get' query requests, or identify specific objects for 'delete', 'cancel' or 'close' requests.
- **Reply:** Required only for response messages to indicate success, failure and error details
- **Payload:** Sometimes required, used to convey message information as a consequence of the 'verb' and 'noun' in the message header

These mentioned elements again contain further elements. The structure of the header itself (Figure 4.6) includes several mandatory attributes like [67]:

- **Verb:** to identify a specific action to be taken. There exists an enumerated set of valid verbs which can be looked up in appendix F.

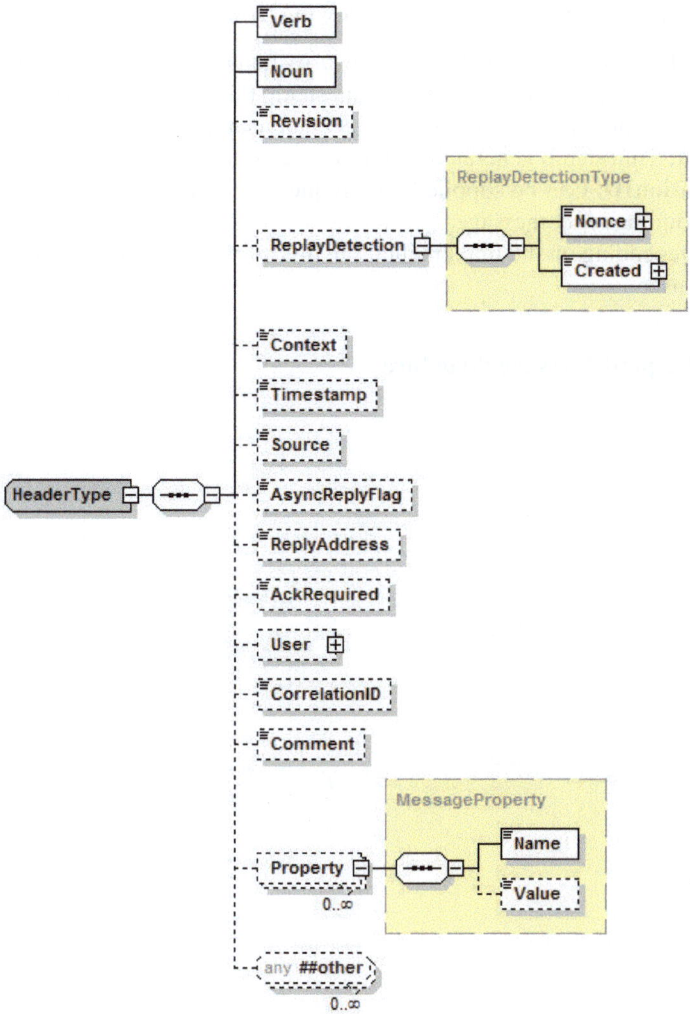

Fig. 4.6 Header message structure [67]

- **Noun:** to identify the subject of the action and the type of payload. In many cases, these are CIM class names with an attached "‘s"’ at the end, such as MeterReadings
- **Source:** to identify the source of a message.
- **Revision:** to indicate the revision of the message definition (in most cases this value is set to one by default)

The optional attributes includes the following [67]:

- **Nonce:** This value is defined by WS-Security and holds a unique number at least for one day. In combination with the timestamp this number can be used to make the message unique and is used to protect against reply attacks
- **Created:** Timestamp generated by the source system
- **CorrelationID:** Can be supplied on a request, so that the client can correlate a corresponding reply message.
- **User:** Depending on the implementation, this attribute could be required. Identifies the user.

4.1.2.1 Request Message Structure

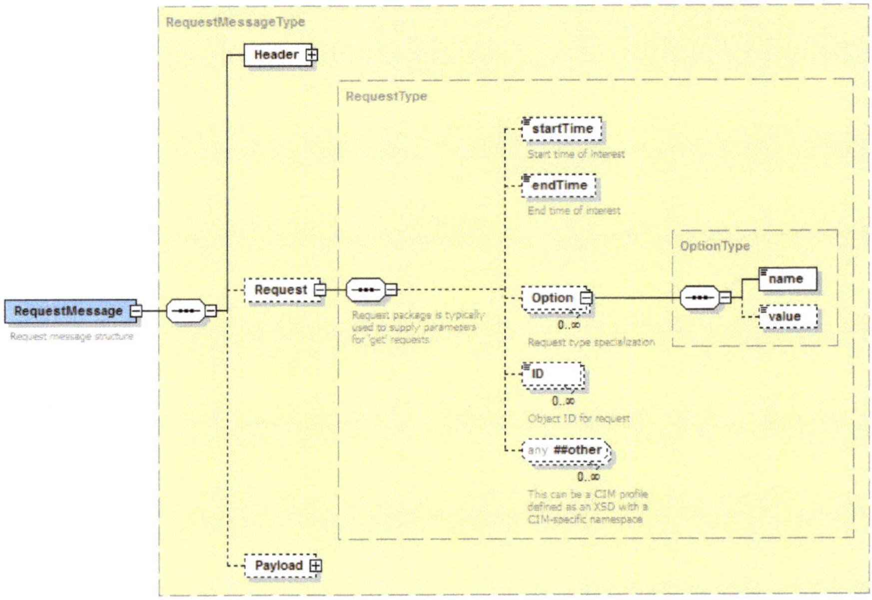

Fig. 4.7 Request message structure [67]

Diagram 4.7 describes the expanded request element for request messages. The given parameters are most commonly used in conjunction with *get* requests as qualifiers. The *StartTime* and *EndTime* are used to specify a time period for the specific request. The *Option Name/Value* pair should only be used with proper discretion for general request options like the following list [67]:

- Specification of a Transaction Timeout Value
- Specifying a Response mode of either "Aggregated" or "Streaming"

Any other use of the *Option*, *Name/Value* pair will be considered as a custom extension to the model. The parameter *ID* can be used in a request messages using the verbs *get*, *cancel* or *close*, to specify a certain object, for example to request information to a specific option. The *any ##other* element can be used to define more complex requests. The following listing shows a sample request message:

```
<ns0:RequestMessage xmlns:ns0 = "http://www.iec.ch/TC57/2008/schema/message">
<ns0:Header>
    <ns0:Verb>get</ns0:Verb>
    <ns0:Noun>Switches</ns0:Noun>
    <ns0:CorrelationID>1729363b5b7d9c6a0a88d02ae97c64b0</ns0:CorrelationID>
</ns0:Header>
<ns0:Request>
    <ns0:ID>35378383838</ns0:ID>
    <ns0:ID>363482488448</ns0:ID>
    <ns0:ID>894094949444</ns0:ID>
</ns0:Request>
</ns0:RequestMessage>
```

The example message represents a *get(Switches)* request for the object ID 35378383838, 363482488448, and 894094949444. A *CorrelationID*with the value: 1729363b5b7d9c6a0a88d02ae97c64b0 is given to be used by the response message, to allow a precise assignment of the incoming messages.

4.1.2.2 Response Message Structure

Figure 4.8 shows the recommend response message structure, which should be used to return the response to the corresponding request message.

The result element can have the three following values graphically presented in 4.9:

1. **OK:** Used in the case there are no errors and all results have been returned
2. **PARTIAL:** Used if only a partial set of has been returned
3. **ERROR:** In case of an error in processing the request

The optional error element could hold further information about specific errors. Figure 4.10 describes the details of the error type. Furthermore, the ID element can be used to reply a transaction ID if needed.

The following listing shows a sample reply message without an error [67]:

```
<ns0:ResponseMessage xmlns:ns0 = "http://www.iec.ch/TC57/2008/schema/message">
<ns0:Header>
    <ns0:Verb>reply</ns0:Verb>
    <ns0:Noun>Switches</ns0:Noun>
    <ns0:CorrelationID>1729363b5b7d9c6a0a88d02ae97c64b0</ns0:CorrelationID>
</ns0:Header>
<ns0:Reply>
    <ns0:ReplyCode>OK</ns0:ReplyCode>
</ns0:Reply>
<ns0:Payload>
    ...
</ns0:Payload>
</ns0:ResponseMessage>
```

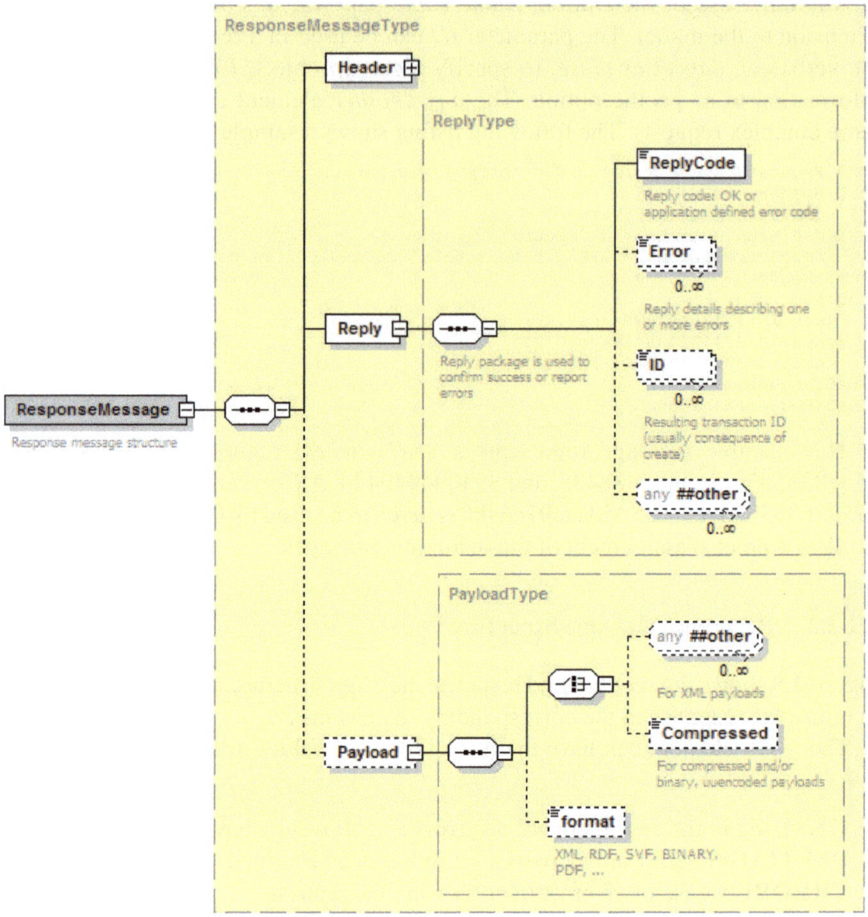

Fig. 4.8 Response message structure [67]

The *reply(Switches)* messages is a response message to a reply message with the **CorrelationID**:1729363b5b7d9c6a0a88d02ae97c64b0. The *ReplyCode* value "OK" indicates that the request could be processed without a problem. The requested information would be part of the payload but is left out in the example.

4.1.2.3 Event Message Structure

An event message is normally used to report a condition to interested systems. The verbs used in an event message are past tense (e.g. created) and will not include either request or reply parameters but only header and usually a payload as shown in figure 4.11.

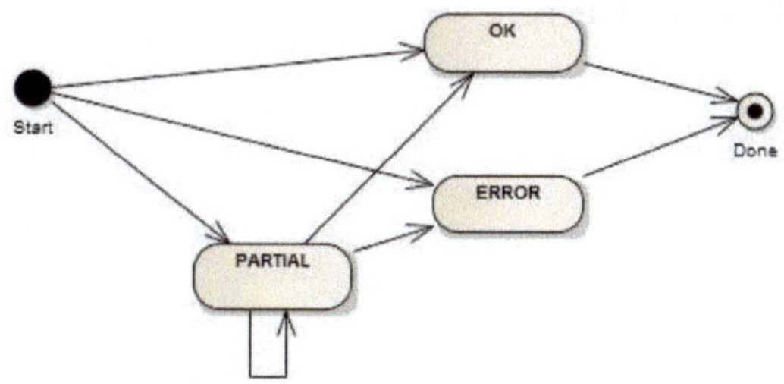

Fig. 4.9 Response message state transition diagram[67]

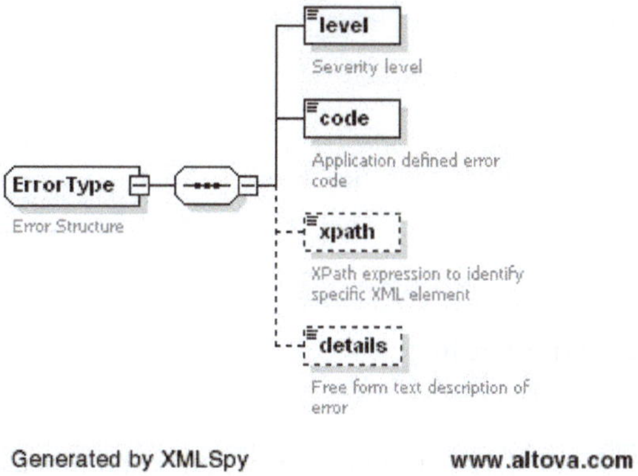

Fig. 4.10 Error type structure[67]

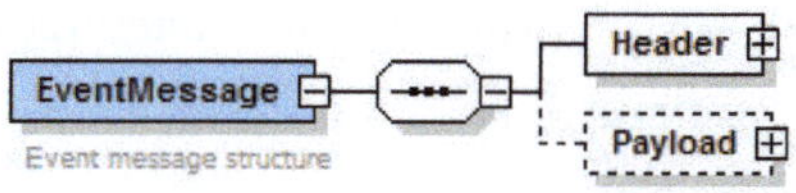

Fig. 4.11 Event message structure [67]

A event message could look like the following listing [67]:

```
<ns0:EventMessage xmlns:ns0 = "http://www.iec.ch/TC57/2008/schema/message">
<ns0:Header>
    <ns0:Verb>changed</ns0:Verb>
    <ns0:Noun>Switches</ns0:Noun>
</ns0:Header>
<ns0:Payload>
    ....
</ns0:Payload>
</ns0:EventMessage>
```

4.1.2.4 Fault Message Structure

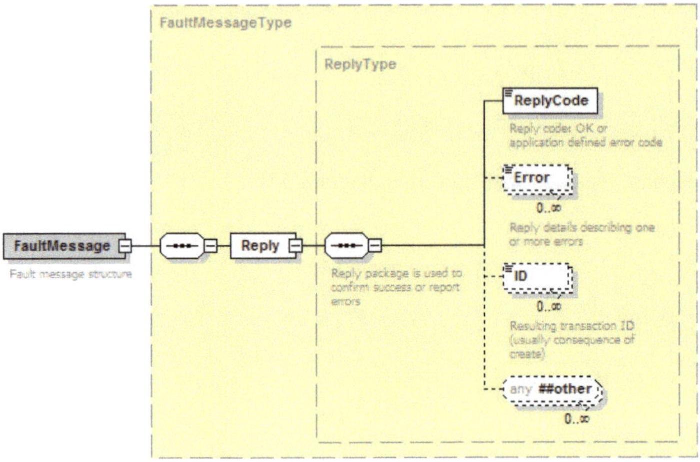

Fig. 4.12 Fault message structure [67]

The fault message (figure 4.12 is used to report fault condition as a consequence of a failed attempt to process a request. A fault message only holds the reply package, as it may not have been able to even interpret the header of a request message. The following listing shows a sample fault message:

```
<ns0:ResponseMessage xmlns:ns0 = "http://www.iec.ch/TC57/2008/schema/message">
<ns0:Reply>
    <ns0:ReplyCode>ERROR</ns0:ReplyCode>
    <ns0:Error>
        <ns0:level>ERROR</ns0:level>
        <ns0:code>222</ns0:code>
        <ns0:details>Unknown object ID: 35378383838</ns0:details>
    </ns0:Error>
</ns0:Reply>
</ns0:ResponseMessage>
```

This message could be a result of a reply message, which requests further information on an object with the *ID*: 35378383838.

4.1.2.5 Payload Structure

The last missing part for a common message is the payload element. There are two forms of payload structures: generic and type-specific. Where the common message envelope defines the payload as being generic (or type-independent), the definition of Web Service Definition Language (WSDL) for web services may require a payload definition of a variant that is type specific [67].

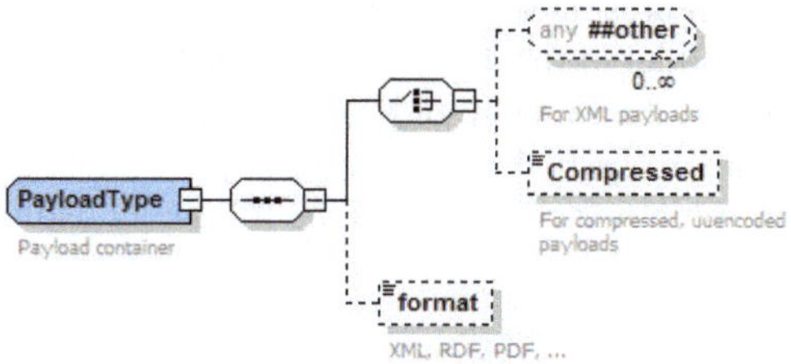

Fig. 4.13 Generic payload structure [67]

The generic payload structure (figure 4.13 includes an *any ##other* element, which holds the XML payload, a *Compressed* element to sign the message if a compression is used, and a *Format* element to describe the payload (like XML, RDF, SVG...). Most payloads contain XML documents that conform to a defined XML schema, exceptionally payloads which holds RDF files or other non-XML formats. Sometimes large payloads have to be compressed minimize bandwidth use.

In case of a type specific payload the structure has to be redefined to be operation specific with a combination of a verb and a noun. The necessary substitution includes the root element *Message* which will be substituted with i.e. "CreatedEndDeviceControl" and the the payload with the concrete message type element such as "EndDeviceControl".

The resulting message diagram can be seen in figure 4.14. The sample names have substituted the generic elements.

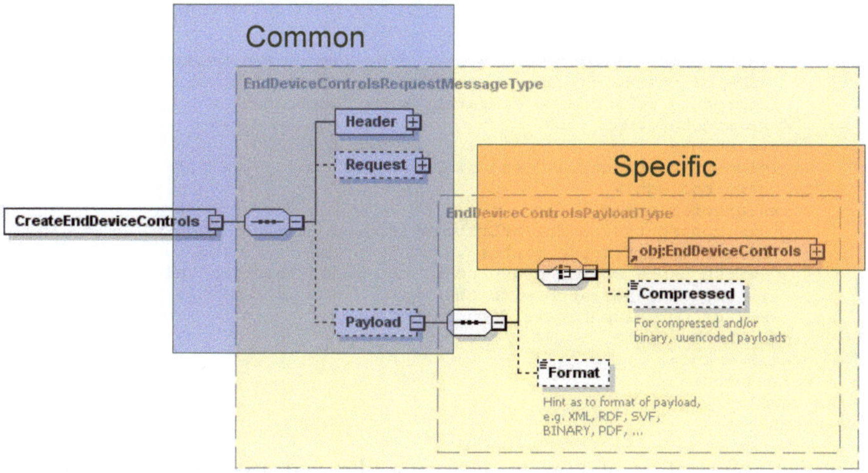

Fig. 4.14 Specific payload structure [67]

Each payload type is represented to a specific noun and are typically derived from the IEC CIM data model in form of design artifacts like XSD. These artifacts are derived from the UML data model and can be seen as profiles of the original data models. The diagram in figure 4.15 shows a sample artifact generated from the UML data model as a profile in XSD whereas the following listing shows the XSD source code:

```
<?xml version="1.0" encoding="utf-8"?>
<xs:schema xmlns:xs="http://www.w3.org/2001/XMLSchema"
xmlns:a="http://www.iec.ch/2008/Message#"
targetNamespace="http://iec.ch/TC57/2008/CIM-schema-cim12#"
elementFormDefault="qualified" attributeFormDefault="unqualified"
xmlns="http://iec.ch/2005/Message#"
xmlns:m="http://iec.ch/TC57/2007/CIM-schema-cim12#">
  <xs:element name="Switches" type="m:Switches" />
  <xs:complexType name="Switches">
    <xs:sequence>
      <xs:element name="Switch" type="m:Switch" minOccurs="0"
      maxOccurs="unbounded" />
    </xs:sequence>
  </xs:complexType>
  <xs:complexType name="Switch">
    <xs:annotation>
      <xs:documentation>A generic device designed to close, or
      open, or both, one or more electric
      circuits.</xs:documentation>
    </xs:annotation>
    <xs:sequence>
      <xs:element name="mRID" minOccurs="1" maxOccurs="1"
      type="xs:string">
        <xs:annotation>
          <xs:documentation>A Model Authority issues mRIDs. Given
          that each Model Authority has a unique id and this id is
          part of the mRID, then the mRID is globally
          unique.</xs:documentation>
        </xs:annotation>
      </xs:element>
      <xs:element name="name" minOccurs="0" maxOccurs="1"
      type="xs:string">
        <xs:annotation>
          <xs:documentation>The name is a free text human readable
          name of the object. It may be non unique and may not
          correlate to a naming hierarchy.</xs:documentation>
        </xs:annotation>
      </xs:element>
      <xs:element name="normalOpen" minOccurs="0" maxOccurs="1"
      type="xs:boolean">
        <xs:annotation>
          <xs:documentation>The attribute is used in cases when no
          Measurement for the status value is present. If the
          Switch has a status measurement the Discrete.normalValue
          is expected to match with the
          Switch.normalOpen.</xs:documentation>
        </xs:annotation>
      </xs:element>
    </xs:sequence>
  </xs:complexType>
</xs:schema>
```

Further information of how to derive profiles and artifacts can be found in chapter 3.8 and of course in the section "Examples of using the CIM" in chapter 5.

The name of the top level element must be the same as the noun that is used in the message header. The *Switch* element as well as the sub level attributes *mRID*, *name*

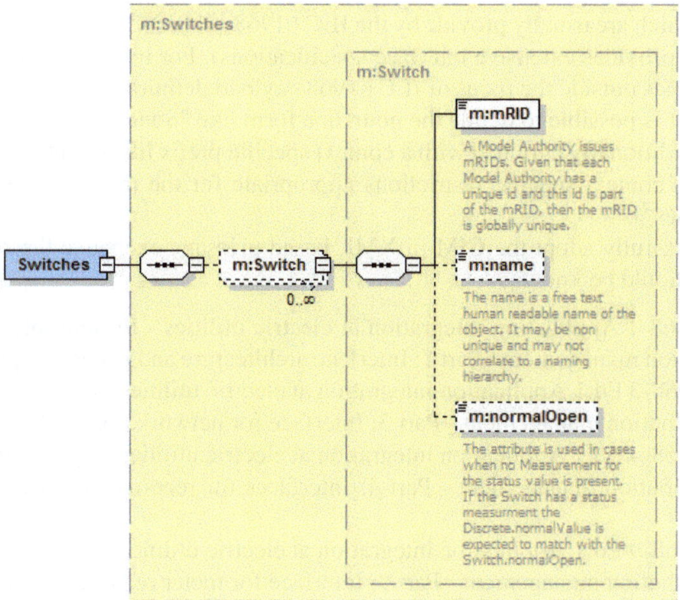

Fig. 4.15 Example switch message payload [67]

and *normalOpen* are derived from the original CIM UML model and are a subset of the attributes hold by the *Switch* object in the data model.

With this XML Schema it is possible to create the following example XML message, which represents three instances of a switch and the corresponding attributes as defined in the schema:

```
<m:Switches
xsi:schemaLocation="http://www.iec.ch/TC57/2008/CIM-schema-cim12# Switches.xsd"
xmlns:m="http://www.iec.ch/TC57/2008/CIM-schema-cim12#"
xmlns:xsi="http://www.w3.org/2001/XMLSchema-instance">
  <m:Switch>
    <m:mRID>35378383838</m:mRID>
    <m:name>SW1</m:name>
    <m:normalOpen>true</m:normalOpen>
  </m:Switch>
  <m:Switch>
    <m:mRID>363482488448</m:mRID>
    <m:name>SW2</m:name>
    <m:normalOpen>true</m:normalOpen>
  </m:Switch>
  <m:Switch>
    <m:mRID>894094949444</m:mRID>
    <m:name>SW3</m:name>
    <m:normalOpen>false</m:normalOpen>
  </m:Switch>
</m:Switches>
```

These specific payloads should be defined by interface specifications using XML Schemas which are usually provide by the IEC 61968 subparts 3 to 9 (at the moment only part 9 provides extensive interface specifications). For implementations using CIM messages outside the focus of IEC 61968 payload definitions can be defined as necessary. It is possible to define the noun in a form like "Switches", "BidSet". The noun can additionally enriched with a context specific prefix like <context><noun>. This allows context specific restrictions appropriate for the message context (i.e. "GetSwitches").

To successfully adopt the CIM in XML based message exchange the following standards should be known:

- IEC 61968-1 Application integration at electric utilities - System interfaces for distribution management - Part 1: Interface architecture and general requirements
- IEC 61968-3 Ed.1 Application integration at electric utilities - System interfaces for distribution management - Part 3: Interface for network operations
- IEC 61968-4 Ed.1 Application integration at electric utilities - System interfaces for distribution management - Part 4: Interfaces for records and asset management
- IEC 61968-9 Ed.1 Application integration at electric utilities - System interfaces for distribution management - Part 9: Interface for meter reading and control
- IEC 61968-14 System Interfaces For Distribution Management -XML Naming and Design Rules
- IEC 61970-1 Ed.1 Energy management system application program interface (EMS-API)- Part 1: Guidelines and general requirements
- IEC 61970-2 Ed.1 Energy management system application program interface (EMS-API) - Part 2: Glossary
- IEC 61970-301 Ed.1 Energy management system application program interface (EMS-API) - Part 301: Common information model (CIM) base
- IEC 61970-402 Ed.1 Energy management system application program interface (EMS-API) - Part 402: Common services

4.2 Exchange of Network Topology Data by Using RDF Serialization

The exchange of network topology data is clearly a very important use case for the CIM. It deals with the important task of exchange of network topologies between different parties and includes necessary physical (like lines, transformers...) and virtual objects(like voltagelevel). To achieve this, the CIM data model has to be serialized in an adequate form. The IEC decided to use the RDF Syntax derived by the W3C, which is further described in 2.3 for the following reasons as the proposed solution should be:

- machine and human readable
- accessible through tools which support Document Object Model(DOM) and other XML application program interfaces

- self describing
- taking advantage of W3C recommendations

The RDF technology can match all of the above requirements. The main serialization from the abstract UML Model to RDF is described in IEC 61970-552-4 "Energy Management System Application Program Interface (EMS-API) - Part 552-4 CIM XML Model Exchange Format" [59], which hast not finally be released up to today.

This standard provides information about the process of exchanging topology models in general as well as the workflow, the basic serialization process and additional information in the modeling process (like creator, version, ...). It also mentions the use of CIM RDF Schema as described in IEC 61970-501 "Common information model resource description framework (CIM RDF) Schema". The general idea is to generate a common basis for the exchange by using the schema, a brief overview can be seen in 4.16.

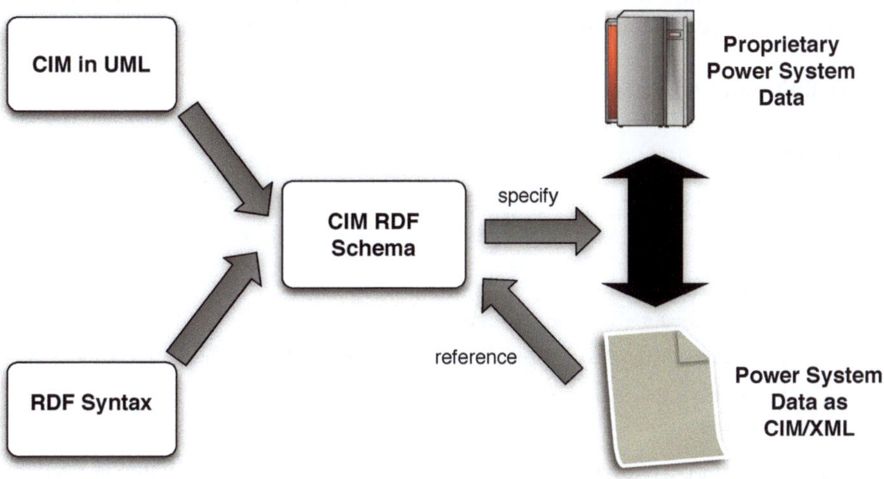

Fig. 4.16 CIM XML based power system model exchange mechanism [38]

Because the workflow and the additional information can slightly differ according to the given scenario, in this chapter we focus on the basic serialization process and introduce the general syntax defined by the standard. The following part describes certain elements and the serialization, which is mainly extracted from the IEC 61970-552-4 standard [59], but does not cover all handled points out of this standard.

Definition of a Definition-Element

Syntax:

```
<classname rdf:ID=identity>
   <!-- Content:
   (literal-property|units-property|resource-property|compound-property)*
   -->
</classname>
```

Example:

```
<cim:PowerTransformer rdf:ID="A_37EFD2AB3">
...
...
</cim:PowerTransformer>
```

The definition element introduces a new resource and gives its type. The element type, classname, is the XML qualified name of a class from the CIM schema. The value of the *ID* attribute, *identity*, is a XML name symbol. It is not necessarily related to the power system resource but it has to be unique in the document.

Definition of a Literal-Property Element

Syntax:

```
<propname>
<!-- Content: text -->
</propname>
```

Example:

```
<cim:PowerTransformer rdf:ID="A_37EFD2AB3">
   <cim:IdentifiedObject.aliasName>Trafo</cim:IdentifiedObject.aliasName>
   <cim:IdentifiedObject.mRID>1</cim:IdentifiedObject.mRID>
</cim:PowerTransformer>
```

The literal-property element introduces a property and a literal value applying to the enclosing resource. The element type, *propname*, is the XML qualified name of a property from the CIM schema declared as a namespace in the document element.

Definition of a Resource-Property Element

Syntax:

```
<propname rdf:resource=resource-uri/>
```

Example URI-Reference:

```
<cim:PowerTransformer rdf:ID="A_37EFD2AB3">
   <cim:PowerTransformer.Contains_TransformerWindings
 rdf:resource="#A_3233F563E"/>
</cim:PowerTransformer>
```

Example Enumeration:

```
<cim:PowerTransformer rdf:ID="A_37EFD2AB3">
  <cim:PowerTransformer.phases rdf:resource=
 "http://iec.ch/TC57/2008/CIM-schema-cim13#PhaseCode.A"/>
</cim:PowerTransformer>
```

The resource-property element introduces a property and a resource as its value applying to the enclosing resource. The element type, *propname*, is the XML qualified name of a property from the CIM schema declared as a namespace in the document element. The resource-uri is an unique URI-reference that identifies a resource.

The above described elements should cover most of the needed serializations.

To maintain the interoperability the following guideline from [59] should be used:

1. Include the likely primary key properties of each resource at the the first time it is introduced. This is to support applications like database applications if they require a primary key, which is most likely the case.
2. Prefer single-valued properties (like Equipment.MemberOf_EquipmentContainer) instead of their many-valued inverse(like EquipmentContainer.Contains_Equipments) to simplify the editing or transforming of the document. Because the properties are inverse, there will be no information loss.
3. Single value relationships should be represented by including only one side in the serialization
4. By specifying an rdf:ID the XML naming conventions should be used available in [126].
5. The use of stable URIs for resources across multiple documents or versions is recommend.

To handle a potentially existing extension of the CIM with new classes and attributes, namespaces can be used (more information on extending the CIM can be found in chapter 3.9. Because of the separate namespaces the customized documents clearly delineate what belongs to the CIM Standard and what is custom. A further explanation of the Namespaces itself can be found in chapter 2.2 . The following namespace declaration identifies an extended namespace with the name "offis":

```
xmlns:offis="http://www.offis.de/cim/extension"
```

To add a non-CIM attribute like "‘registrationNumber"’ to the Powertransformer class, the example might look like the following code-snippet:

```
<cim:PowerTransformer rdf:ID="_37EFD2AB3">
   <cim:IdentifiedObject.mRID>1
 </cim:IdentifiedObject.mRID>
   <offis:registrationNumber>RG44331</offis:registrationNumber>
</cim:PowerTransformer>
```

The RDF schema describing this extension can be separated into a different RDF schema document and therefore can be clearly set aside from the original RDF schema to enable the possibility of independent modifications in each document.

In conclusion, it can be said that knowledge of the following standards are necessary to successfully adopt the CIM in this use case.

- IEC 61968-13 Ed.1 Application integration at electric utilities - System interfaces for distribution management - Part 13: CIM RDF Model exchange format for
- IEC 61970-1 Ed.1 Energy management system application program interface (EMS-API) - Part 1: Guidelines and general requirements
- IEC 61970-2 Glossary
- IEC 61970-301 Ed.2 Energy management system application program interface (EMS-API) - Part 301: Common information model (CIM) base
- IEC 61970-452 Energy Management System Application Program Interface (EMS-API) - Part 452: CIM Transmission Network Model Exchange Profile
- IEC 61970-501 Common information model resource description framework (CIM RDF) Schema
- IEC 61970-552-4 Energy Management System Application Program Interface (EMS-API) - Part 552-4 CIM XML Model Exchange Format distribution.

Part II
The CIM: Application and Examples

Chapter 5
Examples of Using the CIM

5.1 XML Based Message Exchange

The XML based message exchange is one of the main use cases to the application of the CIM. Roughly said, there are two applications available. The first one is the standardized message exchange based on standardized interfaces specified in the standards IEC 61968-3 to IEC 61968-9. The second one is the coupling of two heterogeneous systems with a custom interface.

Basically, the XML message exchange between systems is platform independent, but in the following example the use of an ESB is assumed. Furthermore, the use case does not include the WSDL creation for the Webservices but only the basic XML Schema to generate those.

5.1.1 Standardized Message Exchange

The IEC 61968-9: Application integration at electric utilities - System interfaces for distribution management - Part 9: Interface for meter reading and control [57] is one of the newer standardized parts of the CIM and will be used in this chapter to demonstrate the standardized message exchange.

In this use case, we assume to have a Smart Meter and an Meter Data Management (MDM) System. The MDM system provides a common point of management, data repository and data aggregation for several Smart Meter. These Smart Meters can detect outages and have to report to the MDM system. To implement such a message in CIM, the IEC 61968-9 standard provides the necessary information such as sequence diagrams and the XML Schema to generate an interface. Additionally, the recommended message structure, introduced in chapter 4.1, will be used.

With the aforementioned components provided by the standard, it is quite easy to implement an interface for this use case. The flow diagram (figure 5.1) shows the relevant systems and the message action to implement.

Corresponding systems in this scenario are only the Smart Meter itself and the MDM System. The message should have the name *EndDeviceNames*. Additionally, the verb *Created* will be used.

M. Uslar et al.: The Common Information Model CIM, POWSYS, pp. 127–148.
springerlink.com

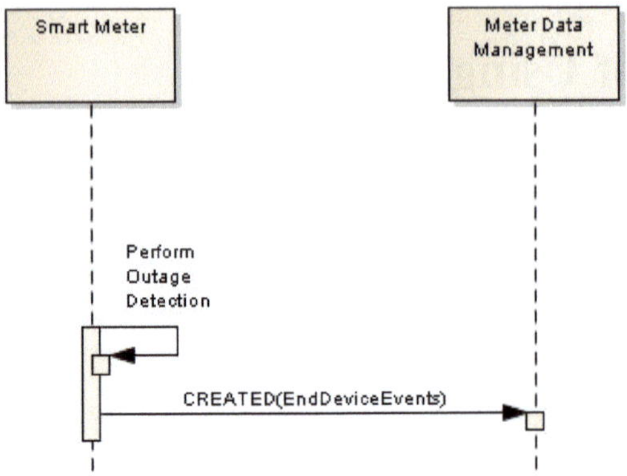

Fig. 5.1 Adjusted sequence diagram of reporting an outage from a Smart Meter [57]

The payload for the message *EndDeviceNames* is already defined in the standard. The graphical representation is shown in figure 5.2. The *EndDeviceEvent.category* element holds the description to the event such as outage, health or power quality. A recommendation of available enumerations for the categories can be found in the annex of the IEC 61968-9 Standard. Additionally, a time stamp (*EndDeviceEvent.createdDateTime*) is required as well as a identification code (*EndDeviceEvent.Asset.mRID*) of the sending asset. (e.g. system).

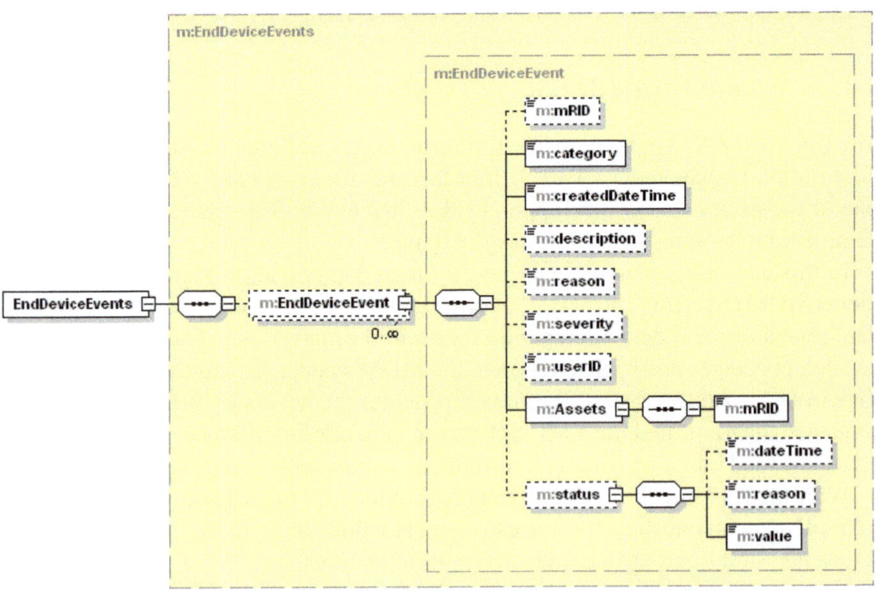

Fig. 5.2 EndDeviceEvents message schema [57]

The XML schema source code to this figure can be found in appendix D. Based on the schema this sample message can be formed to accomplish the requirement of the use case to generate a message to transport an outage message:

```
<?xml version="1.0" encoding="utf-8"?>
<m:EndDeviceEvents xsi:schemaLocation=
        "http://iec.ch/TC57/2009/EndDeviceEvents#example.xsd"
xmlns:m="http://iec.ch/TC57/2009/EndDeviceEvents#"
xmlns:xsi="http://www.w3.org/2001/XMLSchema-instance">
  <m:EndDeviceEvent>
    <m:category>2.36.1.29.</m:category>
    <m:createdDateTime>2011-06-22T09:30:47</m:createdDateTime>
    <m:Assets>
      <m:mRID>7788392ED82EA1D</m:mRID>
    </m:Assets>
  </m:EndDeviceEvent>
</m:EndDeviceEvents>
```

The above message payload represents an alarm event with the time stamp *2011-06-22T09:30:47* from a metering device with the identification code *7788392ED82EA1D*. The category *2.36.1.29.* is build on the aforementioned recommendations of enumerations in the IEC 61968-9 standard and indicates a failure of the internal clock.

Please note: This practice of putting hidden semantics into instances, as it is done with the alarm category, is not approved by everyone in the standardization and CIM community.

The message payload structure has now to be integrated into the recommended message structure shown in chapter 4.1.2 and the XML Schema in appendix B. The graphical representation of the complete interface specification as XML schema is shown in figure 5.3. The resulting source code to the schema can be found in appendix D.

This artifact can now be used by middleware specific tools to implement the interface and can be used to exchange messages regarding end device events.

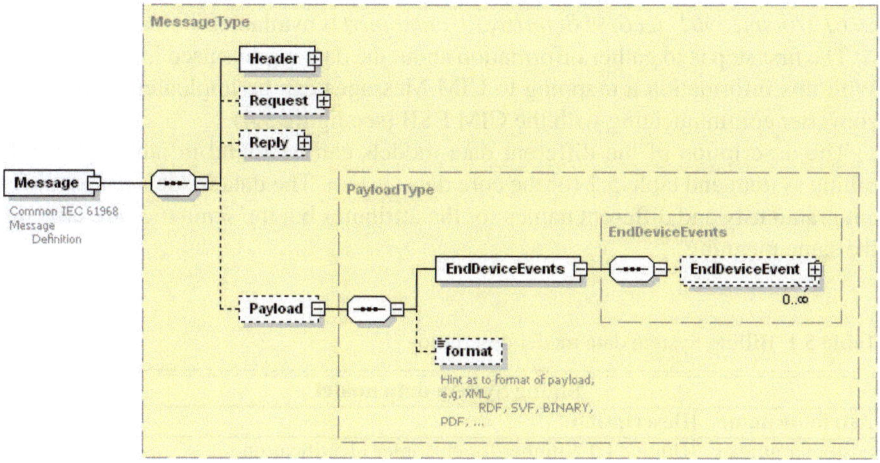

Fig. 5.3 EndDeviceEvents complete message schema [57]

5.1.2 Coupling of Heterogeneous Systems

This use case describes the coupling of two ore more heterogeneous systems with the help of CIM based messages. Because of the specific needs there is no such standard as the IEC 61968-9 standard mentioned before for this use case. More steps have to be done to accomplish this task.

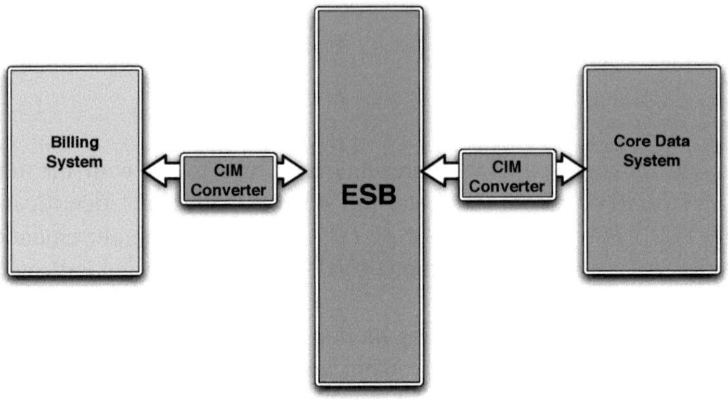

Fig. 5.4 Coupling of two heterogeneous systems

We assume the following: The use case contains two heterogeneous systems, Billing System and Core Data System. The Billing System has to create an invoice for specific users, but only holds one information element about the customer which is the customer ID. The corresponding addresses to this ID is stored in the Core Data System. The systems have different data models and are not directly supporting CIM communication, but a CIM ESB with a CIM data model version *iec61970cim15v01_iec61968cim10v30_combined* is available.

The first step is to gather information about the data models used in both systems. With this information a mapping to CIM Messages can be implemented in a CIM converter communicating with the CIM ESB (see figure 5.4).

The description of the different data models can be found in table 5.1 for the billing system and table 5.2 for the core data system. The data models have a different granularity and different names for the attributes but the semantics are targeting the same meaning.

Table 5.1 Billing System data model description

Billing System data model	
Attribute name	**Description**
customernumber	Unique ID Number for customer identification
adress	Postal address of the customer to send invoices for example

Table 5.2 Core Data data model description

Core Data data model	
Attribute name	**Description**
customerid	Customer identification number (unique)
firstname	First name of the customer
lastname	Last name of the customer
street	street name and street number of the customer
zipcode	post code of the customer
city	city of the customer

To enable a communication between these systems a mapping is needed and (in our use case) both data models should be mapped to CIM. Figure 5.5 shows a selected class and the corresponding data type (compound) from the CIM UML data model, which seems to fit the requirement to hold all the different attributes without the loss of information and is semantically correct.

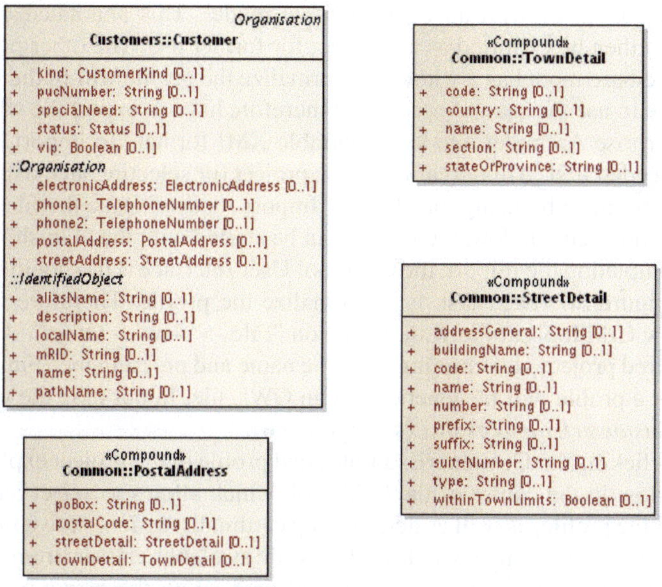

Fig. 5.5 CIM Customer Class

The attributes from the tables 5.1 and 5.2 should be mapped to the selected CIM class. These mapping will represent the internal conversion in the CIM converter as shown in figure 5.4. The mapping necessary can be found in figure 5.6.

In this case, the selected CIM parts only includes one main CIM class and the corresponding data types, but generally the selection can be seen as the process to create a CIM profile.

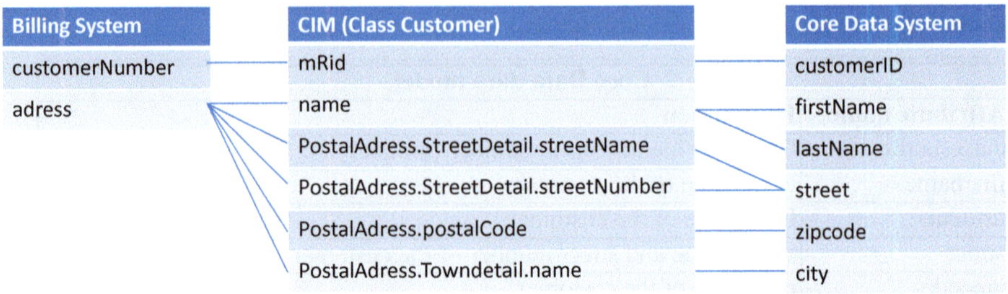

Fig. 5.6 Mapping between data models

At this point the use of tools is recommend. CIMTool offers a wide range of needed features for CIM messaging, therefore the next steps will be explained using the CIMTool.

At first, we have to formalize the selection of the used CIM classes into a profile. For this task there are two possibilities worth mentioning. The first is to formalize the profile in Enterprise Architect (see chapter 6.3.1 with a tool like CIMConteXtor (see chapter 6.2.2) or with an own modified model. This annotated or modified model could then be exported as an artifact for further use. The other possibility is to export the base model as a whole and formalize the profile within the CIMTool.

We chose to use the second option and therefore have to export the whole model out of Enterprise Architect into an importable XMI format. To import the XMI, a project is needed first. After creating a new project (or selecting an existing one) an import can be made by using the "File ⇒ Import" option. Once the file is selected in the new appeared window, the import can be started with the "Finish" button.

After completing the import, the CIMTool User Interface (UI) should look something like figure 5.7. Next task is to formalize the profile. Therefore we need to create a new CIMTool profile with the action "File -> New -> Other". After selecting the desired project, determining a unique name and pressing the "Finish" button, the base for a profile will be generated as an OWL file. In our case the name of the profile is *CustomerExamples.owl* (see figure 5.8).

Double clicking the CustomerExamples.owl profile in the project explorer, opens a new window in the center of the CIMTool, which allows to select and add new elements to the profile, as well as determining cardinality as it is shown in figure 5.9. The right part of the center window shows all available classes from the model, whereas the left window shows the selected ones. With the arrow buttons at the bottom further elements can be added to the profile or removed again.

The far-right window shows the outline of the profile and provides an hierarchical overview. To add properties (attributes and associated classes) to the classes into the profile, the class has to be selected in the outline. After selecting the class, the available attributes and associations are shown in the middle window and can be added to the profile in the same way as the classes before with the arrow buttons at the bottom.

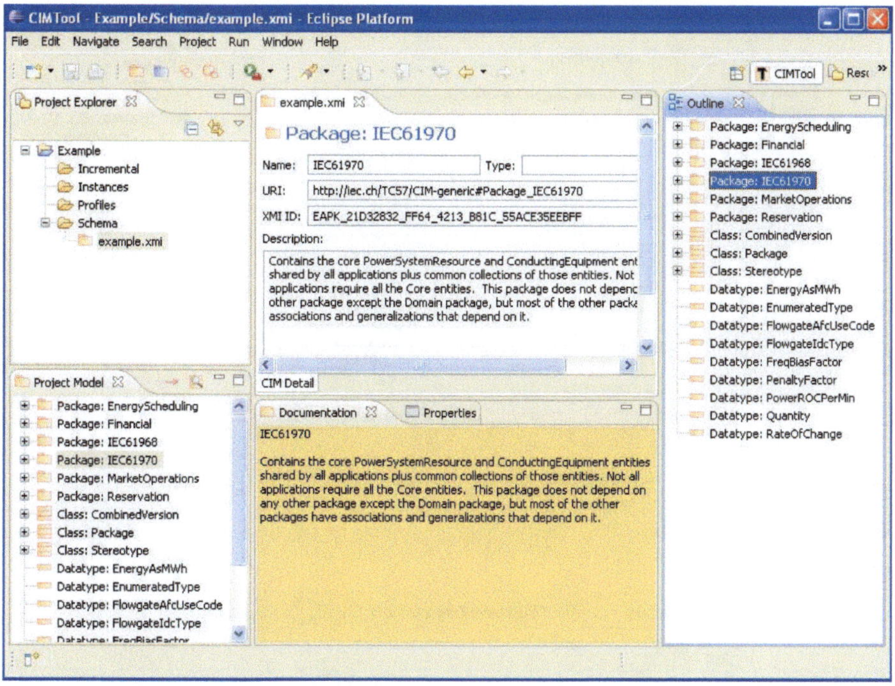

Fig. 5.7 CIMTool after importing the CIM data model

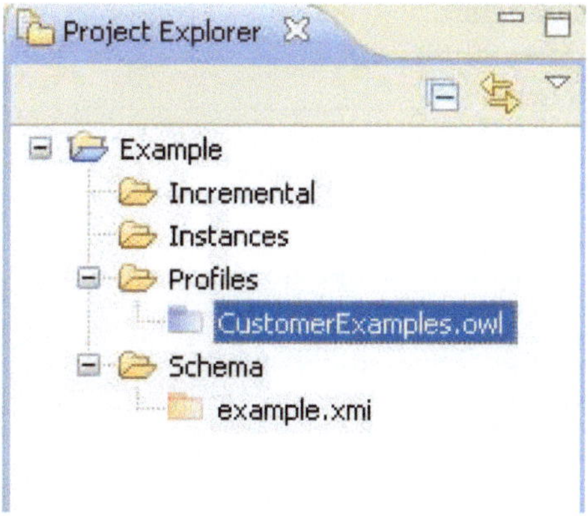

Fig. 5.8 Project Explorer view on CustomerExamples profile

To complete the profile, the definitions of cardinalities have to be done. In this case we do not change the default settings of a cardinality of *1...1* because they fit the requirements.

The formalized profile can now be used inside the CIMTool to generate an XML schema specifically for this profile. This can be done by selecting the tab **Summary**

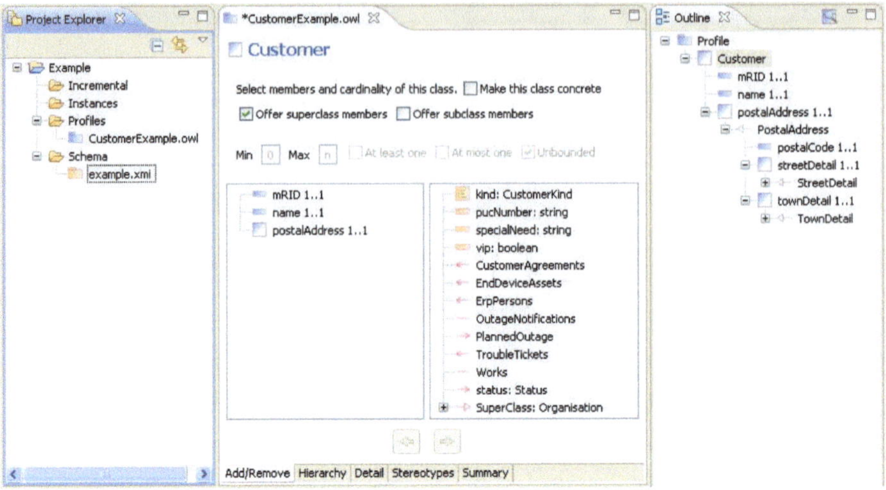

Fig. 5.9 Profile creation

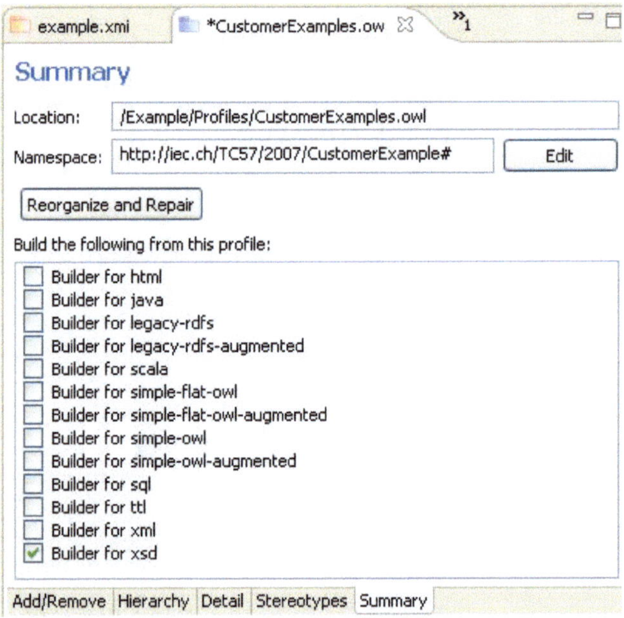

Fig. 5.10 XML schema creation

in the center window at the bottom. On this page, a summary of formats, which CIMTool can generate is shown. To create the XML schema the last option *Builder for xsd* has to be selected (shown in figure 5.10). After saving the project the schema will be build in the background and saved in .xsd file format. The file name is the same as the profile name, in our case *CustomerExample.xsd*.

Figure 5.11 shows the generated schema in a graphical form whereas the source code in XML Schema can be found in appendix C.

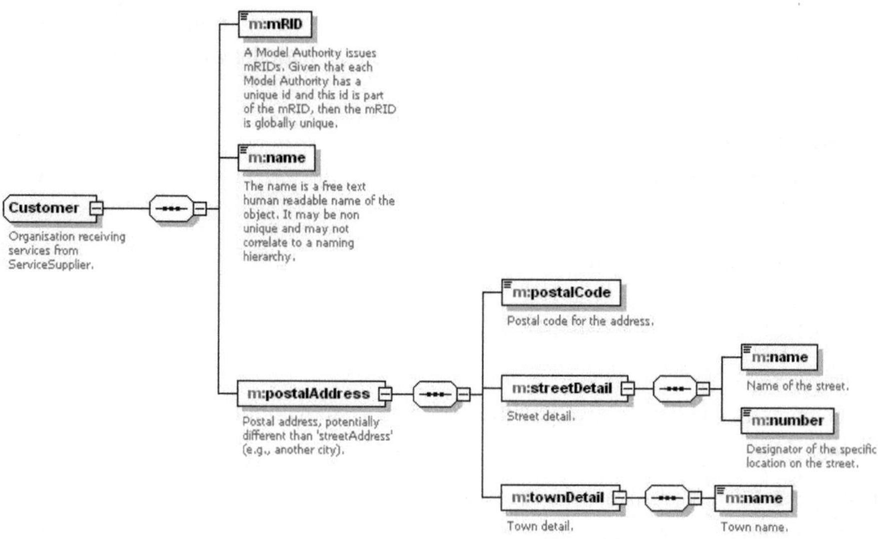

Fig. 5.11 Resulting schema

This artifact can now be used with middle ware specific tools to generate the desired interfaces. The converter have to be build separately.

To demonstrate a scenario, the sequence in figure 5.12 will be used. The billing system (through the connected converter) sends the request message to the ESB

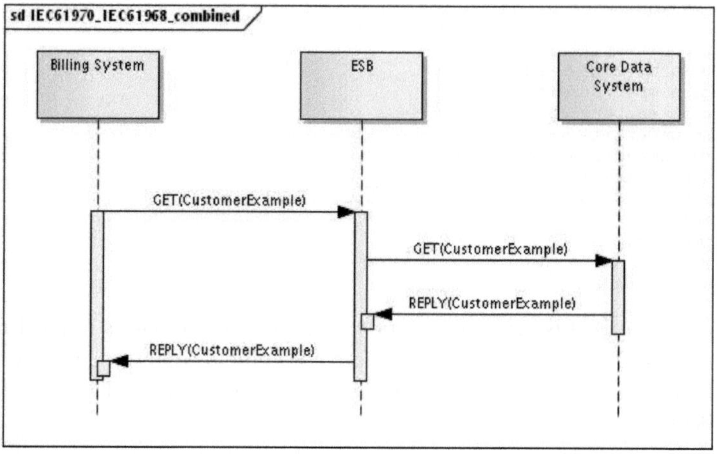

Fig. 5.12 Sequence diagram for the example use case

using the verb *get* and the noun *CustomerExample*. The ESB will then forward the message to the core data system (through the connected converter), which processes the request and replies with a message using the verb *reply* and the same noun.

Messages for a example scenario can look like the following listings (based on chapter 4.1.2. The example request message requests information regarding the customer id *79866654756*. A CorrelationID has been set with the value *17da97aec64b0*, additionally.

Listing 5.1 Example request message

```
<ns0:RequestMessage xmlns:ns0 = "http://www.iec.ch/TC57/2008/schema/message">
<ns0:Header>
    <ns0:Verb>get</ns0:Verb>
    <ns0:Noun>CustomerExample</ns0:Noun>
    <ns0:CorrelationID>17da97aec64b0</ns0:CorrelationID>
</ns0:Header>
<ns0:Request>
    <ns0:ID>79866654756</ns0:ID>
</ns0:Request>
</ns0:RequestMessage>
```

The message response given by the core data system could look like the following listing:

Listing 5.2 Example response message

```
<ns0:ResponseMessage xmlns:ns0 = "http://www.iec.ch/TC57/2008/schema/message">
<ns0:Header>
    <ns0:Verb>reply</ns0:Verb>
    <ns0:Noun>CustomerExample</ns0:Noun>
    <ns0:CorrelationID>17da97aec64b0</ns0:CorrelationID>
</ns0:Header>
<ns0:Reply>
    <ns0:ReplyCode>OK</ns0:ReplyCode>
</ns0:Reply>
<ns0:Payload>
<m:CustomerExamples xmlns:m="http://iec.ch/TC57/2007/CustomerExample#">
  <m:Customer>
    <m:mRID>79866654756</m:mRID>
    <m:name>John Smith</m:name>
    <m:postalAddress>
      <m:postalCode>26121</m:postalCode>
      <m:streetDetail>
        <m:name>Escherweg</m:name>
        <m:number>2</m:number>
      </m:streetDetail>
      <m:townDetail>
        <m:name>Oldenburg</m:name>
      </m:townDetail>
    </m:postalAddress>
  </m:Customer>
</m:CustomerExamples>
</ns0:Payload>
</ns0:ResponseMessage>
```

The *CorrelationID* is set to the same value as at the previous request message. The *ReplyCode* OK indicates that there were no problems regarding the request. The payload holds the needed address information about the requested customer with the id 79866654756.

With this the requirements are fulfilled and the billing system is capable of generating the bills.

5.2 Exchange of Network Topology Data by Using RDF Serialization

5.2.1 Topology Representation

This chapter explains the generation of CIM/XML based topology representation on a specific example. In this example we assume that an existing network shall be transformed into a CIM/XML topology. To achieve this quite a few steps are necessary and will be described here.

1. **Analyze existing data**
   It is inevitable to analyze the existing data of the source topology. This includes the data about electrical properties as well as meta data and semantically connected data. Graphical information for example single line diagrams should be sighted.
   These information are necessary to get an overview about all available data which should be used for the transformation.

2. **Create/Select a CIM profile**
   The aforementioned extracted information from the analysis should now be used to create or select the desired CIM profile. As shown in chapter 3.8 several different profiles exists and are used in various surroundings. The option to select or to create a new profile is depending on the specific use case, the state of systems which have to communicate, the internal as well as external requirements, the present regulatory standards and of course it is depending on the data which should be transfered. The preferred serialization of the profile is UML.

3. **Create algorithm**
   Based on the previous steps an algorithm to transform the source data into the CIM/XML RDF topology given by the individual CIM Profile has to be created.

4. **Verify the resulting RDF**
   The resulting RDF topology has to be verified semantically and syntactically as well as the overall completeness. Various tools can be helpful for this task. The basic RDF syntax can be checked with Altova Semantic Works [2] or the online W3C RDF Validator [129]. CIMSpy (chapter 6.2.3) is another tool to check your RDF topology.

To simplify this use case, the data analysis from step 1 will be skipped and only the results will be used for the next step. The example does not cover the necessary data for a full load-flow calculation but should only show how to convert existing data into a CIM/XML RDF topology.

The data analysis of the example in this chapter results in a single line diagram as shown in figure 5.13 and two tables with technical data about the lines and the power transformer represented in the diagram. In this use case, we are going to create a very simple CIM profile based on the existing data to fulfill step 2. To achieve this we will have to find the suitable CIM classes and attributes for this profile as follows.

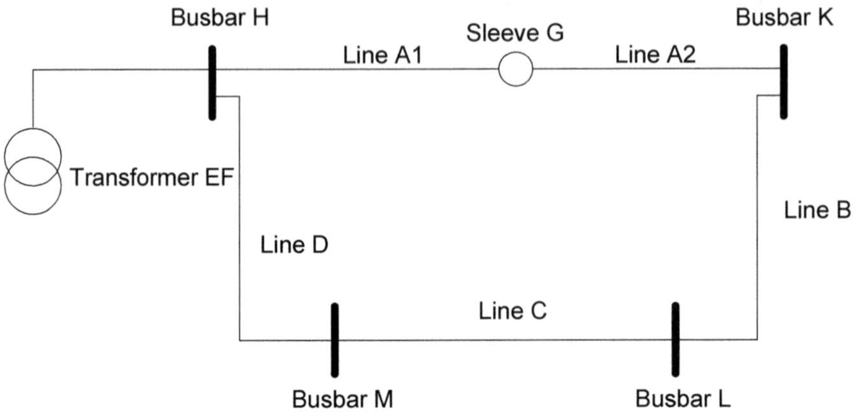

Fig. 5.13 Network topology example

Table 5.3 Result of the analysis of existing data regarding power transformer

Transformer technical data	
Name	**Parameter**
Name	Transformer EF
Id	2
Sr	40 MVA
Winding Connection	YY
u1	110 kV
u2	20 kV

Table 5.4 Result of the analysis of existing data regarding power lines

Line technical data		
Name	**Length in km**	**Parameter**
Line A1 and A2	2.5	r=0.125 Ohm/km
Line B	3.0	x=0.112 Ohm/km
Line C	0.3	c=300nF/km
Line D	2.0	

Power Transformer

The power transformer is a physical object and should therefore be found in the wires package in the IEC61970 part of the CIM UML model. A transformer consists of more than one class to model it completely. According the technical data of the power transformer from table 5.3, several attributes have to be mapped. The CIM defines a power transformer as a container class with an association to the class *TransformerWinding*, which holds the necessary attributes to model the technical data. The mapping of the attributes should be as shown in tables 5.5 and 5.6.

Table 5.5 Mapping from transformer technical data to CIM on class level

Mapping power transformer		
Technical data	**CIM**	**description**
Power Transformer	PowerTransformer	An electrical device consisting of two or more coupled windings, with or without a magnetic core, for introducing mutual coupling between electric circuits. Transformers can be used to control voltage and phase shift (active power flow)
	TransformerWinding	A winding is associated with each defined terminal of a transformer (or phase shifter).

Table 5.6 Mapping from transformer technical data to CIM on attribute level

Mapping power transformer		
Technical data	**CIM**	**CIM description**
Name	PowerTransformer.name	The name is a free text human readable name of the object. It may be non unique and may not correlate to a naming hierarchy
Id	PowerTransformer.mRID	A Model Authority issues mRIDs. Given that each Model Authority has a unique id and this id is part of the mRID, then the mRID is globally unique
Sr	TransformerWinding.ratedS	The normal apparent power rating for the winding
Winding Connection	TransformerWinding.connectionType	The type of connection of the winding
u1/u2	TransformwerWinding.ratedU	The rated voltage (phase-to-phase) of the winding, usually the same as the neutral voltage

The diagram 5.14 shows the original classes out of the CIM UML model whereas snippet 5.15 shows only the relevant classes and attributes. This snippet will be part of the profile for this use case.

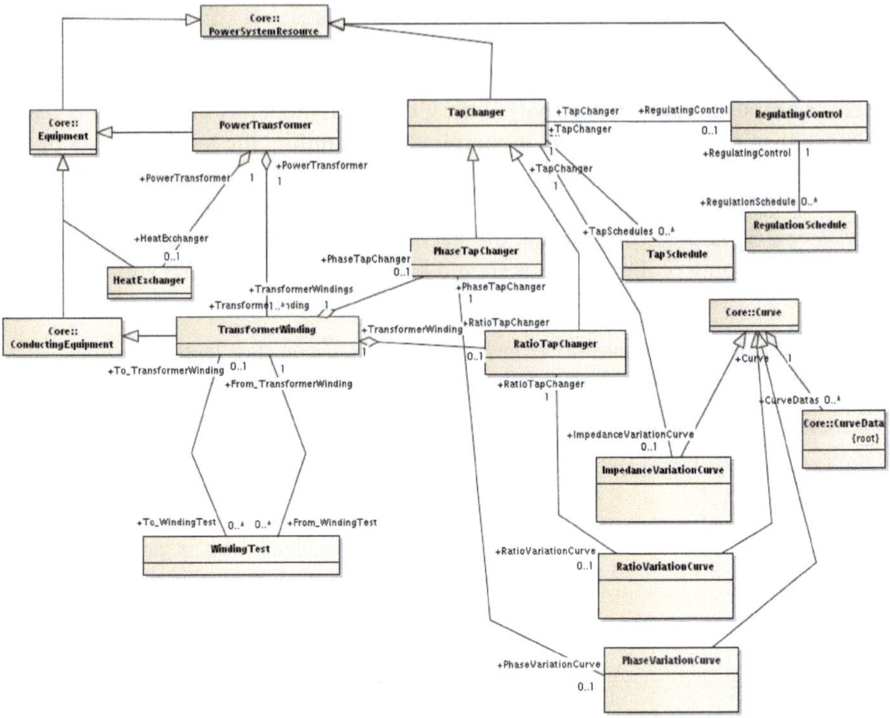

Fig. 5.14 Original CIM power transformer class diagram

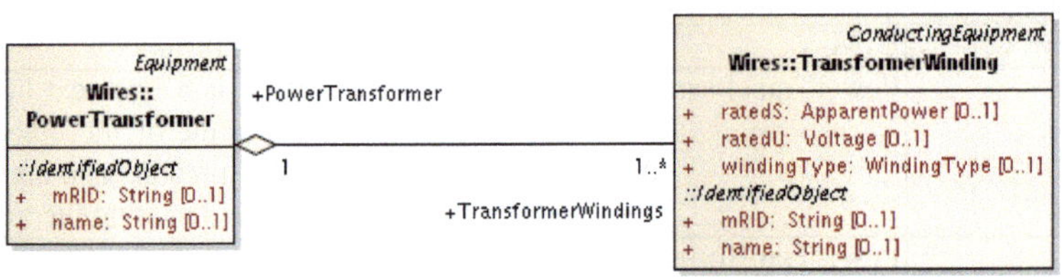

Fig. 5.15 CIM power transformer class diagramm relevant parts

Line

Lines are found in the wires package in the IEC61970 part of the CIM UML model as well. 5.16 shows an overview regarding all line specific classes. The given network example, represents an AC grid, which results in the choice of the class *ACLineSegment* in the model. In addition, the container class *Line* will be needed, to express the state of the lines A1 and A2 representing one line but consisting out of two segments. The class and attribute mapping should be as shown in the tables 5.7 and 5.8.

Table 5.7 Mapping from Line technical data to CIM on class level

Mapping lines		
Technical data	**CIM**	**description**
Line Container(for A1 and A2)	Line	Contains equipment beyond a substation belonging to a power transmission line.
Line	ACLineSegment	A wire or combination of wires, with consistent electrical characteristics, building a single electrical system, used to carry alternating current between points in the power system.

Table 5.8 Mapping from Line technical data to CIM on attribute level

Mapping lines		
Technical data	**CIM**	**CIM description**
Name	ACLineSegment.name	The name is a free text human readable name of the object. It may be non unique and may not correlate to a naming hierarchy
r	ACLineSegment.r	Positive sequence series resistance of the entire line section
x	ACLineSegment.x	Positive sequence series reactance of the entire line section
c	ACLineSegment.bch	Positive sequence shunt (charging) susceptance, uniformly distributed, of the entire line section. This value represents the full charging over the full length of the line

Figure 5.16 shows the whole UML class model, whereas figure 5.17 shows only the relevant parts, which will be used as input for our profile.

Busbar

Compared to the objects line and power transformer, the busbar does not have any technical data in our example. However the object itself can be mapped with the CIM class *BusBarSection* with only an id and a name (see fig. 5.18).

Additional CIM Specific Objects

The previous listed objects power transformer, line and busbar cover all physical objects expressed in this example as shown in figure 5.13. However, to model a CIM topology additional non-physical objects are needed. To link the physical objects together, the CIM uses so called connectivity nodes and terminals. Each object which inherits from *ConductingEquipment* can associate with the class *Terminal*. These *Terminals* again are associated with *ConnectivityNodes* and therefore a connection between different objects can be formed. Figure 5.19 shows the example enriched

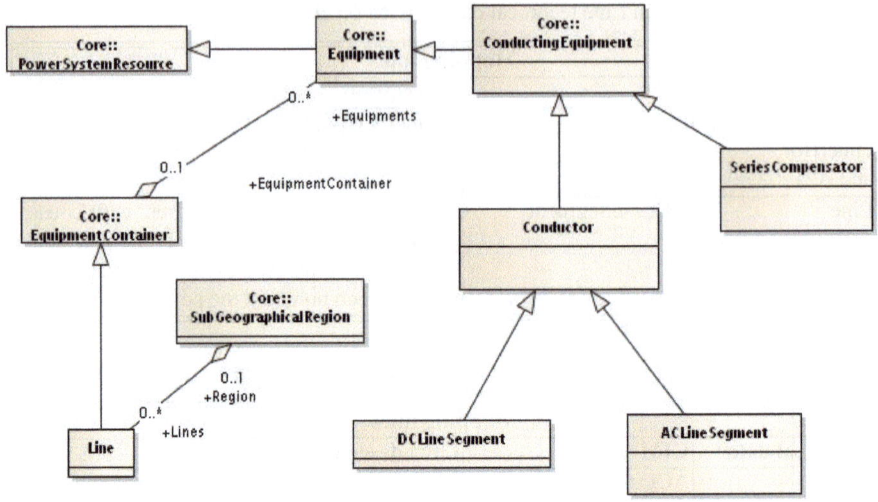

Fig. 5.16 Original CIM line class diagram

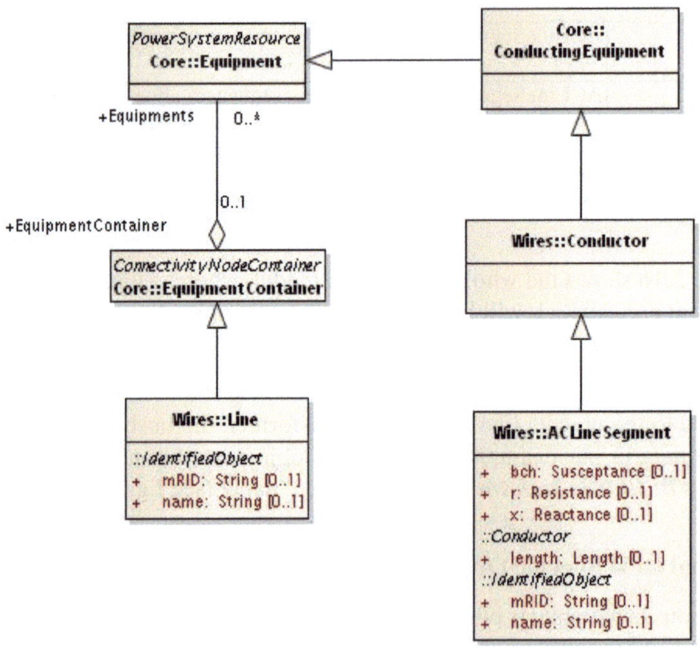

Fig. 5.17 Line class diagram relevant parts

with the non-physical CIM objects and is therefore a complete CIM topology. The used classes are shown in figure 5.20.

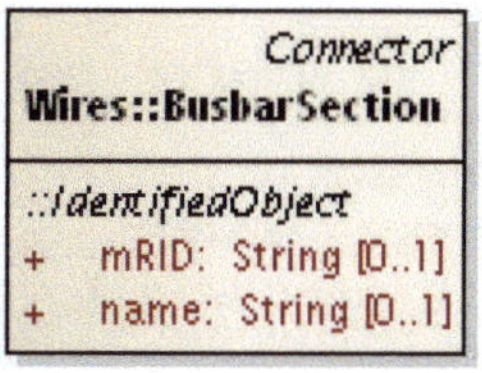

Fig. 5.18 BusBarSection class diagram

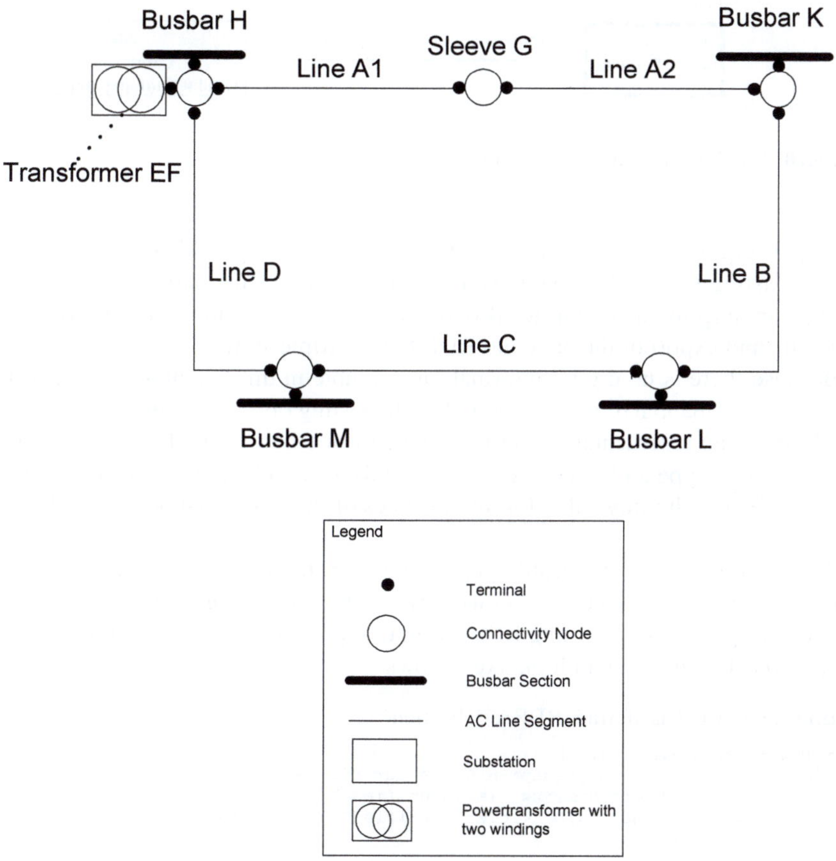

Fig. 5.19 Network topology example transformed to CIM

The profile for this special example can now be put together out of the previous selected classes and attributes. An option is to model this profile in UML as it is shown in figure 5.21. This UML serialization offers the possibility to generate different artifacts for further use, such as RDF or XSD profiles, which can be used to create or validate instances of a topology.

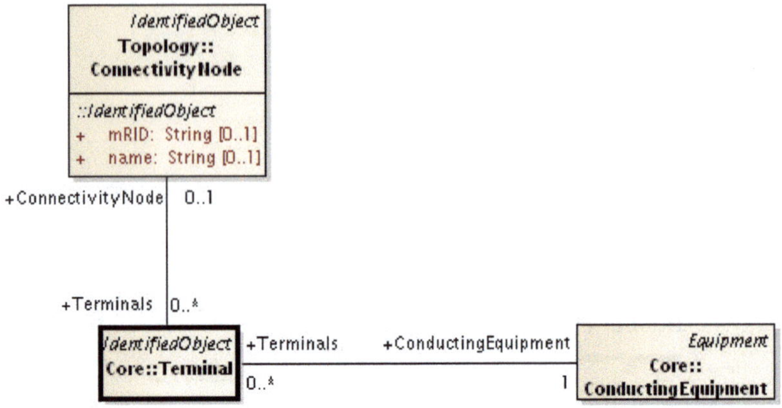

Fig. 5.20 Topology specific class diagram

Step 3 requires the creation of an algorithm to generate CIM/XML RDF based on the existing source data. This includes the reading of the data out of relational database management system (RDBMS) or files, transformation to a CIM compliant structure, and export of the data into RDF files or triple stores.

Because there is no digital information available in this example, the algorithm will result in a manual transformation of each existing object in figure 5.19 resulting in a RDF string representation. At first, we describe how to model the RDF string for each object type and of course we use the RDF serialization rules and tips from chapter 4.2. The dummy value for all attributes of the individual objects will be set to "x".

As we know out of the profile generating step, the power transformer consists of the two different object types *PowerTransformer* and *TransformerWinding*. According our profile the resulting RDF string to represent a *PowerTransformer* and a *TransformerWinding* would look like follows:

Listing 5.3 PowerTransformer RDF serialization

```
<cim:PowerTransformer rdf:ID="x">
  <cim:Equipment.MemberOf_EquipmentContainer rdf:resource="#x"/>
  <cim:IdentifiedObject.name>x</cim:IdentifiedObject.name>
  <cim:IdentifiedObject.mRID>x</cim:IdentifiedObject.mRID>
</cim:PowerTransformer>
```

Listing 5.4 TransformerWinding RDF serialization

```
<cim:TransformerWinding rdf:ID="x">
  <cim:TransformerWinding.connectionType x"/>
  <cim:TransformerWinding.ratedS>x</cim:TransformerWinding.ratedS>
  <cim:TransformerWinding.ratedU>x</cim:TransformerWinding.ratedU>
  <cim:TransformerWinding.windingType x"/>
  <cim:TransformerWinding.MemberOf_PowerTransformer rdf:resource="x"/>
  <cim:IdentifiedObject.mRID>x</cim:IdentifiedObject.mRID>
</cim:TransformerWinding>
```

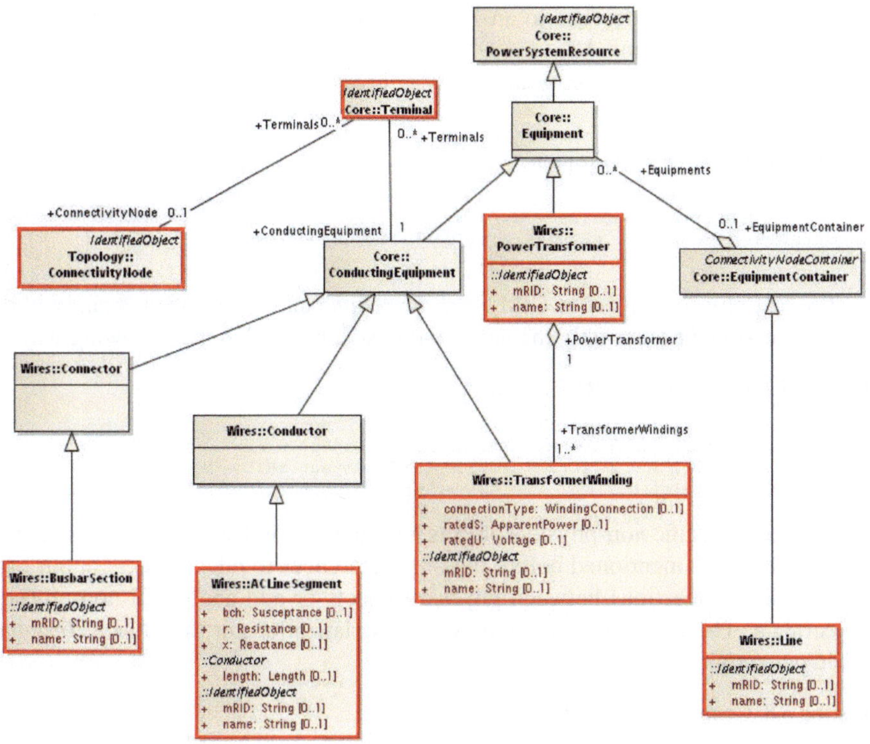

Fig. 5.21 CIM Profile

As for the mapping, there are only two special conversions necessary between source and target values. First the *connectionType* attribute has to be split for each winding. The source data describes the connection type as "YY", so each winding gets only one character. Additionally, a *windingType* can be specified which helps to distinguish between the primary and the secondary winding.

The RDF string representing an ACLineSegment is as follows:

Listing 5.5 ACLineSegment RDF serialization

```
<cim:ACLineSegment rdf:ID="x">
  <cim:Conductor.length>x</cim:Conductor.length>
  <cim:Conductor.r>x</cim:Conductor.r>
  <cim:Conductor.x>x</cim:Conductor.x>
  <cim:Conductor.bch>x</cim:Conductor.bch>
  <cim:Equipment.MemberOf_EquipmentContainer rdf:resource="x"/>
  <cim:IdentifiedObject.name>x</cim:IdentifiedObject.name>
  <cim:IdentifiedObject.mRID>x</cim:IdentifiedObject.mRID>
</cim:ACLineSegment>
```

At this point, a few conversions are needed to match the source and target values. First of all, the technical data for r,x and c are per km. In the CIM class *ACLineSegment* the values should be for the entire line (see mapping table reftable:linemapping2). The capacity *c* value has to be converted into conductance *bch*, if necessary, a reference to an EquipmentContainer has to be created, in this case for the lines A1 and A2.

The container class *Line*, without any conversion needed, can be serialized as follows:

Listing 5.6 Line RDF serialization

```
<cim:Line rdf:ID="x">
    <cim:IdentifiedObject.name>x</cim:IdentifiedObject.name>
  <cim:IdentifiedObject.mRID>x</cim:IdentifiedObject.mRID>
</cim:Line>
```

The last physical object in CIM *BusBarSection* can be serialized, again without any conversion, because only the name and id will be used in the following way:

Listing 5.7 BusBarSection RDF serialization

```
<cim:BusbarSection rdf:ID="x">
 <cim:IdentifiedObject.name>x</cim:IdentifiedObject.name>
 <cim:IdentifiedObject.mRID>x</cim:IdentifiedObject.mRID>
</cim:BusbarSection>
```

Topology specific non-physical objects have to be serialized as well. There are no data attributes as mentioned before but only references to other objects in our case to establish connections between physical objects or add additional non-physical object data. The following listing shows the serialization for the objects *Terminal* and *ConnectivityNode*:

Listing 5.8 ConnectivityNode RDF serialization

```
<cim:ConnectivityNode rdf:ID="x">
</cim:ConnectivityNode>
```

Listing 5.9 ConnectivityNode RDF serialization

```
<cim:Terminal rdf:ID="x">
  <cim:Terminal.ConductingEquipment rdf:resource="x" />
  <cim:Terminal.ConnectivityNode rdf:resource="x" />
</cim:Terminal>
```

We now have a serialization of each existing object of the example. The only work left is to fill the dummy value "x" with its real value for each attribute, and of course to fill in the right references to connected objects. These reference are not based on existing data, but can be generated within the creation process. The connection at "Sleeve G" in picture 5.19 consists of two *Terminal* and one *ConnectivityNode* and with the references set it should be like the following listing:

Listing 5.10 ConnectivityNode RDF serialization

```
<cim:Terminal rdf:ID="A_T5">
  <cim:Terminal.ConductingEquipment rdf:resource="#A_A1" />
  <cim:Terminal.ConnectivityNode rdf:resource="#A_CN2" />
</cim:Terminal>
<cim:ConnectivityNode rdf:ID="A_CN2">
</cim:ConnectivityNode>
<cim:Terminal rdf:ID="A_T6">
  <cim:Terminal.ConductingEquipment rdf:resource="#A_A2" />
  <cim:Terminal.ConnectivityNode rdf:resource="#A_CN2" />
</cim:Terminal>
```

These references have to be set at each object, to complete the RDF graph structure. With this done, the topology CIM/XML RDF file is complete. The entire file is printed in appendix E.

The last step for generating an accurate and thorough topology document, is to verify the results so far. The syntax can be checked with classical RDF tools like the W3C RDF Validator (available at [129]) or desktop software like Altova Semantic Works [2]. The outcome of verifying the example with this tools shows no syntactic errors in the document.

It is insufficient to test the RDF syntax alone. Tools like CIMSpy SE 6.2.4 can be used to check whether the RDF document is valid according to a selected profile. The CIMSpy freeware version has a built-in rule set included which can be used for basic testing.

Note: The resulting RDF file of this example will throw errors in CIMSpy, because we made our own profile and not all mandatory associations and attributes which are in the CIMSpy SE rule set have been applied to it.

5.2.2 Difference Model Exchange

As CIM does not offer an official data-base model and data exchange is based on the RDF representation of CIM (though called CIM/XML), it seems consequential to use an RDF-based database instead of a relational-database schema. This section is going to discuss the CIM and its special application from the perspective of RDF-based systems [10]. The main use cases when dealing with a topology exchange are as follows, we identified several use cases for the exchange of topology information for power grids between energy distribution companies:

- **Exchange of complete topologies at regular intervals:** Hitherto, data is exchanged using proprietary formats which leads to high complexity. Adapters have to be implemented and maintained. Due to this, it is only feasible to exchange data on a quarterly-hour basis. Additional time and effort is spent on validation of complete topologies by experts, where only the changes from the last data exchange are of interest.
- **Exchange of parts of topologies:** In order to be able exchange parts of topologies, i.e. voltage levels, substations, a flexible mechanism is needed to extract required information from topologies.
- **Merging of independently created topology parts:** The merging of multiple, independently created topology parts into one consistent topology is a major issue for the collaboration between utilities. Imagine a scenario where multiple regional utilities cooperate with a grid provider. The grid provider needs to merge the topologies provided by the regional utilities into one consistent topology in order to evaluate it. This process raises questions regarding responsibility for topology parts and consistency.
- **Traceability of topology evolutions:** Power grids are constantly evolving and changing their state. These changes must be saved and made accessible.

In order to implement these use cases, the model developed by Britton on regional model authorities [3] proves useful. The regional model for dealing with CIM topologies was originally developed by Britton as a theoretical model. It is used to evaluate the responsibilities and interfaces between different utilities having to interconnect their grids while still adhering to requirements from their own infrastructures. We have to distinguish between three types of regions according to the research by Britton [3]. The first region, the base region, is directly under control and responsibility of utility. The so called boundary region, the second region, forms a border between different base regions. All changes within a boundary region have to be validated by the dispatchers in charge of the associated base regions. Resources from one region may only have associations to resources belonging to the same region or to associated base regions. Aggregated regions, our last type of region, are defined being as the aggregation of a number of base regions with their respective boundary regions. In a system, the topologies are stored and responsibilities for certain base regions are assigned to different users. They are held responsible for their individual base region and have to communicate with the other operators when they have to commit a change either to a boundary region or another base region.

It is inefficient to transfer complete serialized topologies. To overcome this inefficiency, a difference format for CIM-topologies needs to be developed. DeVos [10] proposed a difference format for the CIM. Unfortunately, this format has some drawbacks. It does not adhere to the current W3C RDF standard and extends the standard with a new parseType. It uses a syntax with insertions and deletions which might be unintuitive for users, because most changes to topologies are modifications, resulting in one insertion and one deletion. Additionally, deVos only specifies a format, but no method to compute the difference between two RDF-graphs. A solution and possible system implementation to cope with those issues is presented in [114].

Chapter 6
Toolsupport

6.1 General Introduction and Overview

The use of tools to integrate the CIM is inevitable to minimize the implementation effort and the costs. Also the tools can minimize the error rate and ensure further interoperability if the same tool chains are used, so generated artifacts (like RDF Schema or XML Schema files) as partial result can be exchanged between different parties.

Table 6.1 shows the most common and accepted tools in the CIM community. Some tools have been discontinued in the last years (i.e. CIMPhony) and some other tools are only used for model or standard maintaining and are left out at this point. The information about the tools were provided at a CIM User Group meeting and are mostly taken from [14] or corresponding websites.

Of course, there are several other tools which can be used which are not specialized for the CIM domain. The CIM reuses well-known technologies, which have wide spread support from tools already. For XML based technologies tools like Altova XML Spy or Oxygen are well known tools. RDF and OWL are more special but have been established in the last years. Protégé and Altova Semantic Works are notable tools in this context. For each used technology different tools exist which can be used in the context of CIM.

M. Uslar et al.: The Common Information Model CIM, POWSYS, pp. 149–167.
springerlink.com

Table 6.1 Overview on common tools supporting the integration of CIM (derived from [14])

Name	**Purpose**	**Platform**	**License**	**Link**
CIMBench	Develop and generate CIM-based XML schema for message payloads	Dedicated software	COTS	Contact www.btc-ag.com
CimConteXtor	CIM UML model-based tool to transform UML model to message schemas	Enterprise Architect Add-in	Freeware	www.cimcontextor.net
CIMTool	CIM profile and message schema creation and validation	Eclipse plug-in	Open Source	www.cimtool.org
CIMSpy	CIM/RDF/XML file browser, editor, and validator	DHTML client application	GNU public licenses	www.powerinfo.us
CIM EA	CIM modeling and design	Enterprise Architect Add-in	Freeware	www.cimea.org
Enterprise Architect	Dedicated Software	UML model management and maintenance	Licensed for fee	www.sparxsystems.com
CIMDesk	Web based application	CIMDesk is a special edition of CIMSpy built for the purpose of ENTSO-e network requirements	Licensed for fee	www.powerinfo.us

6.2 Open Source Tools

6.2.1 CIMBench

Name	BTC CIM Bench
Purpose	Develop and generate CIM-based XML schema for message payloads & model APIs
Platform	Dedicated software, Microsoft .NET / Silverlight
Short Description	• import CIM models via XMI • development of CIM-based XML schema • browsing, exploring and searching the CIM • use of extension models • generate different APIs • translating the CIM • comparing CIM models
Link	`www.btc-ag.com`

The BTC CIM Bench is the all-purpose CIM development tool by the BTC AG. It was designed for the modeling of CIM messages and to aid in the exploration of the CIM. It grew out of CIM-related software projects, starting as a tool for internal use, developed by programmers for programmers to assist in the enterprise application integration with CIM.

6.2.1.1 Features

The BTC CIM Bench was presented at the 2010 CIM users group meeting in Milan. The BTC CIM Bench supports XMI as a UML exchange format, enabling it to load and display all CIM versions exported by Enterprise Architect or Rational Rose. The BTC CIM Bench is used to explore and search the CIM, to provide translations (currently only German is available), to construct a message model and to generate schemas for XML messages. Further, different versions of the CIM can be loaded and compared. The BTC CIM Bench can generate APIs for different programming languages. For the development of message schemas, the header employed by the BTC is included.

6.2.1.2 Application

The BTC CIM Bench is being developed at the BTC AG as a .NET stand-alone application. It is the central tool for all CIM-related activities at BTC. One of the most important benefits of the BTC CIM Bench is the support it gives the user for assembling CIM message models.

After the selected CIM version has been loaded from an XMI file, the BTC CIM Bench will enter a message mode and display the entire CIM model as a tree structure (as shown in figure 6.1). The user can now select the desired classes and assemble a message tree with drag-and-drop. The class structure is displayed with colored symbols so that inherited and native attributes and associations can easily be picked out. The modeling is supported by a search function which allows the searching for classes, attributes, associations and the searching of the documentation. The result of a search is displayed in a window as a clickable list. The translation mode offers a near complete translation into German.

The message tree displays the message model with the selected classes and all their inherited attributes and associations. The multiplicity of attributes can be set to required or optional, while the multiplicity of associations can be chosen freely. To keep the design lean, associations or attributes which are not actually needed for the particular message can be deleted. Associations are only used if they are filled in by a class from the CIM model tree (via drag-and-drop).

In case the CIM proves insufficient for the desired message model, an extension model developed in Enterprise Architect can be loaded from an exported XMI file. Further, the CIM model loaded in the BTC CIM Bench can be restricted by the additional loading of profiles. After the message model has been completely specified, different XML message schemas can be generated. The message models can be saved in project files for future use.

HTML templates are supplied for generating documentation of the message schemas. Additional information regarding the message models, say namespaces or version used, can be edited in a properties window. The generated message schemas can be employed directly for the interfaces in an ESB environment or be used as a basis for generating proxy classes.

The model-driven approach of the BTC CIM Bench allows not only the generating of message schemas from the CIM but also, for instance, APIs for different programming languages. At the moment, templates for generating C++ and C# APIs are available. A complete API of all classes of the CIM version 14 contains 933 classes. The scope of the generated APIs can be restricted through a profile, lest the model must always be considered in its entirety.

In order to assess the differences between two versions of the CIM, a comparison function is provided. After two versions of the CIM have been loaded, differences between them are highlighted with the colors red, yellow and green and are additionally outputted to a text field. The color scheme allows the user to take in the extent of the differences through one quick scan of the model (trees).

The BTC CIM Bench is the central tool for the CIM-based EAI at BTC and has been used successfully - and was actively developed - in several projects. With the first release of the BTC CIM Accelerator for Microsoft's BizTalk Server in August 2011, a Silverlight version of the BTC CIM-Bench, residing in the Windows Azure cloud, will be easily accessible in a web browser.

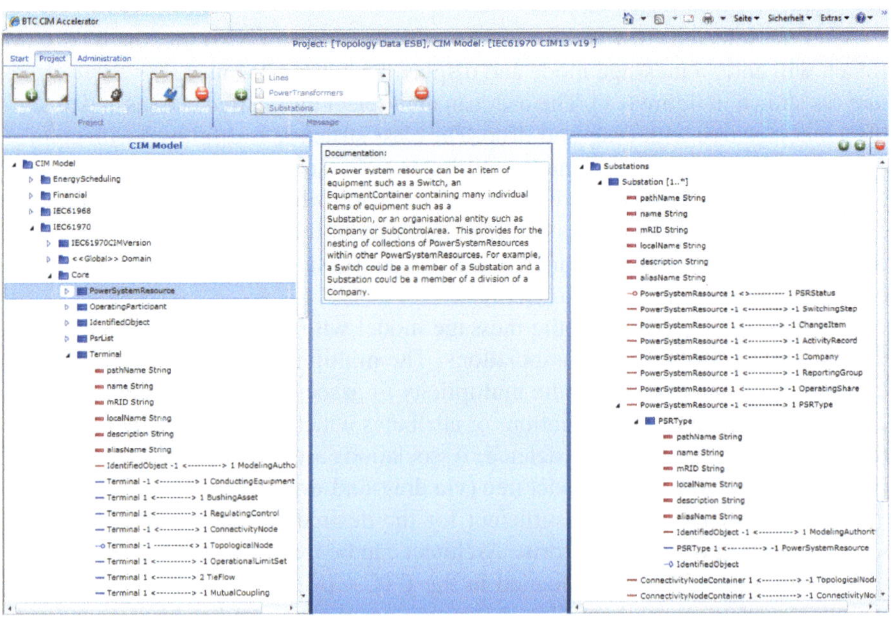

Fig. 6.1 BTC CIM Bench overview

6.2.2 *CimConteXtor and CimSyntaxGen*

Name	CimConteXtor
Purpose	UML model-based tool to create UML profiles
Platform	Enterprise Architect Add-in
Short Description	Follows UN/CEFACT methodology with naming and design rules to create and edit UML profiles
Link	www.cimcontextor.net

Name	CimSyntaxGen
Purpose	CimConteXtor companion add-in to transform UML class diagramm models to message schemas
Platform	Enterprise Architect Add-in
Short Description	Follows UN/CEFACT methodology with layering from information model to contexts to create RDFS and XSD for message/file schemas as well as HTML for documentation purposes
Link	www.cimcontextor.net

CimConteXtor and CimSyntaxGen are both Sparx Systems Enterprise Architect (EA) open source Add-Ins originally provided by Zamiren, a company of André Maizener.

CimConteXtor supports UML profile creation and was developed by Sébastien Maligue-Clausse, André Maizener and Jean-Luc Sanson. The last two developers are involved for years in the CIM development and are both members of the IEC TC 57 WG 13. First, CimConteXtor was developed to build CIM profiles but now it can be used with any UML class diagram model.

Apart from CimConteXtor André Maizener and Jean-Luc Sanson developed a companion Add-In for CimConteXtor called CimSyntaxGen. CimSyntaxGen supports the export of models into syntactic schemas like RDFS, XSD and HTML.

6.2.2.1 Features

CimConteXtor supports UML profile creation based on naming rules as defined by the UN/CEFACT Core Component Technical Specification (CCTS) [89]. CimSyntaxGen provides as a companion tool for CimConteXtor the generation of schemas according to IEC standard like TC 57 XML NDR draft or TC 57 WG 13 IEC 61970 part 501 (RDFS) [88].

Both Add-Ins are based on the CCTS methodology and define four modeling levels [88]:

1. The information model level: contains a domain model like the CIM
2. The profile model level: defines the use of the domain model (or a subset/profile) in a context like the Common Power System Model (CPSM) in the context of power network exchange. Within CimConteXtor the concept "‘IsBasedOn"’ is used.

3. The assembly model level: comprises the definition of how the profile artifacts are assembled for exchange.
4. The syntax model level: specifies the syntax which will be used for exchange and the mapping of the assembly to the selected syntax.

Figure 6.2 shows the parts addressed by CimConteXtor and CimSyntaxGen in relation to the CCTS modeling levels.

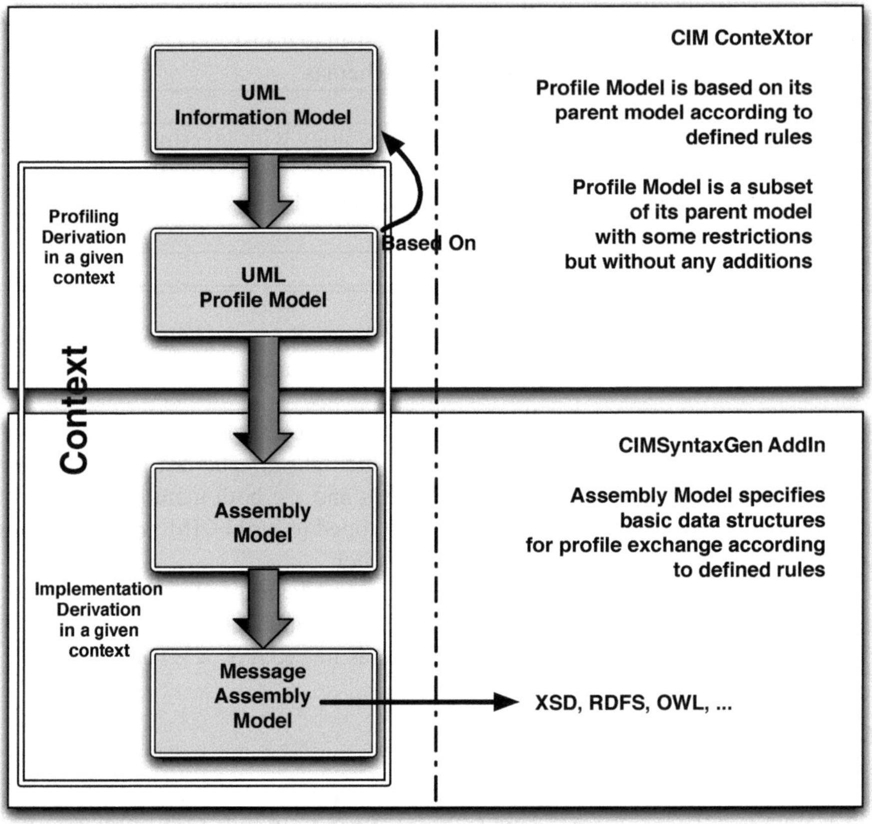

Fig. 6.2 CimConteXtor and CimSyntaxGen overview [88]

Altogether, they provide a CIM UML model-based tool that support the building of profiles and the transformation of UML class diagram models to message schemas. Further CimConteXtor companion add-ins are planned to provide more exports to other syntactic models, like an OWL export [88].

6.2.2.2 Application

Acquisition of both Add-Ins is free but requires registration at www.cimcontextor.net. CimConteXtor and CimSyntaxGen are available as binary distribution for Windows operation systems XP or Windows 7 (Windows Vista is currently not supported) and require Sparx Systems Enterprise Architect version 7.

After installation, the add-ins are accessible through EAs "Add-Ins" tab, see Fig. 6.3. In Fig. 6.3 the XSD exports provided by CimSyntaxGen are shown.

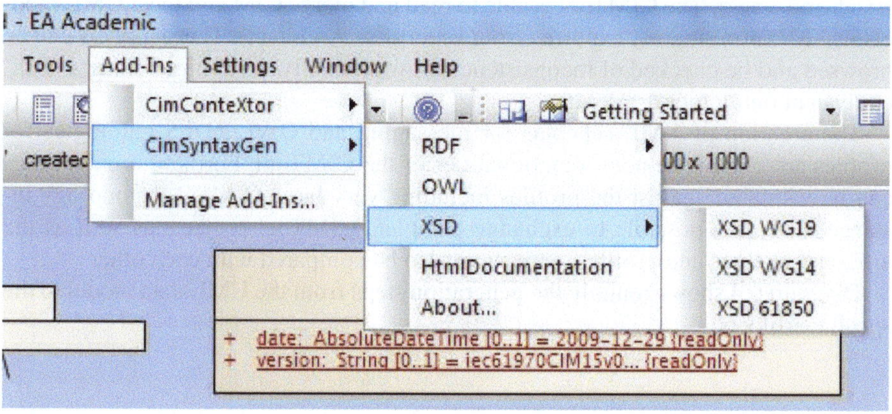

Fig. 6.3 CimConteXtor and CimSyntaxGen EA add-ins

Detailed description on how to use CimConteXtor and CimSyntaxGen are provided in the corresponding user guides on www.cimcontextor.net, see [89] and [88].

6.2.3 CIMTool

Name	CIMTool
Purpose	CIM profile and message schema creation and validation
Platform	Eclipse plug-in
Short Description	• read and merge CIM and local UML models in XMI form • browse models and check inconsistencies • generate equivalent OWL ontologies • create and edit profiles • generate XML schemas, OWL and RDFS ontologies for profiles • validate instances against profiles (including very large CIM/XML instances)
Link	www.cimtool.org

The CIMTool is a general purpose tool for developing and managing CIM-derived models, profiles, and schemas. The tool itself is implemented as a plugin for the eclipse platform and is developed in a joint venture by vendors and utilities. One special point is that CIMTool uses Ontologies in OWL format to define the profiles and mapping models. It is free to use and can be customized if desired.

6.2.3.1 Features

The following feature description are taken from the official webpage `www.cimtool.org`. The CIMTool is able to read and merge CIM and local UML models in XMI form (mainly exported from Enterprise Architect). These models can be browsed and be checked of inconsistencies. Additionally, form these models OWL equivalent ontologies are created.

The creation of XML schemas for messaging and OWL/RDFS ontologies for profiles are one of the main functionalities of the CIMTool. Furthermore instances can be validated against the profiles including very large CIM/XML topology instance files. It is possible to exchange complete CIMTool projects as well as the artifacts schemas and profiles, which can also be compared with each other.

Diagram 6.4 shows roughly the generation steps from the UML data model to the specific artifacts.

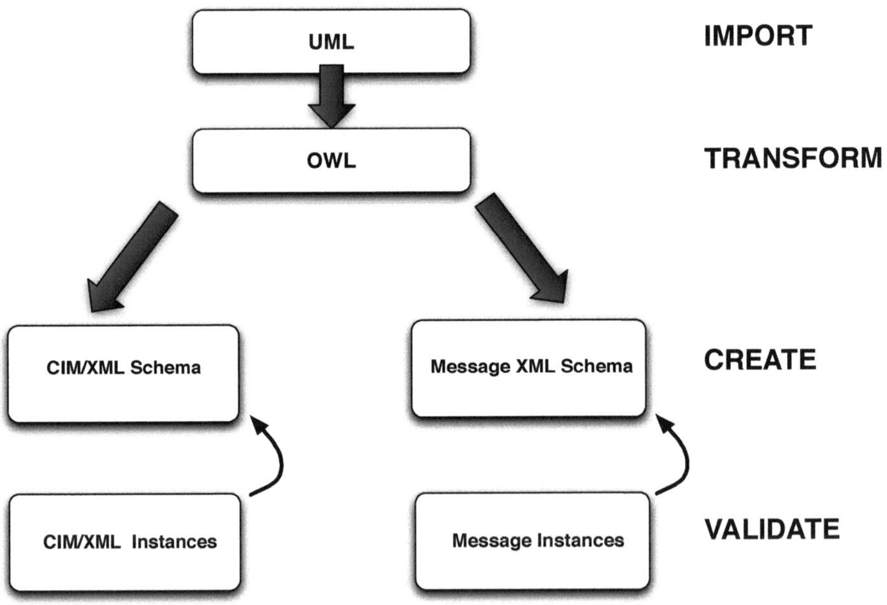

Fig. 6.4 Workflow in CIMTool

6.2.3.2 Application

Because the CIMTool application is already described in the use cases in chapter 5, this section will be left out.

6.2.4 CIMSpy SE

Name	CIMSpy
Purpose	CIM/RDF/XML file browser, editor, and validator
Platform	Internet Explorer 5.0 or above
Short Description	CIMSpy is a CIM-based data engineering tool designed to support CIM-based model exchanging. It provides a wide range of CIM-based data engineering functionalities, including CIM browsing, editing, validation, merging/partition, comparison, and visualization.
Link	www.powerinfo.us

CIMSpy can be described as a CIM based data engineering tool implemented as a Dynamic HTML (DHTML) browser application. The tool has been used in the CIM community for quite a few years and did perform tests in the CIM IOP testing rounds. The tool provides browsing, editing, validation, merging, comparison of CIM data. There are already inbuilt rule-sets implemented for validating against CPSM and domain specific modeling rules.

CIMSpy SE is the open source version of the CIMSpy EE commercial tool, which provides more functionality. The tool CIMDesk is build on the CIMSpy EE tool and is further described in chapter 6.3.2.

6.2.4.1 Features

The latest feature list can be found at the Power Info web site [101]. CIMTool makes it possible to explore different CIM/XML documents, through transforming the XML documents into HTML web pages, including animations, sorting and organizing in hierarchical views.

It is possible to navigate through a model or to search specific elements by strings. The integrated validators are offering the possibility to validated CIM/XML models against CIM/CPSM schema or the rule based validator can be used to detect domain specific modeling errors.

6.2.4.2 Application

The tool itself can be downloaded from `www.powerinfo.us` and the installation process only consists of extract the file to the local file system. To start the application the *CIMSpy.htm* has to be opened with the Microsoft Internet Explorer.

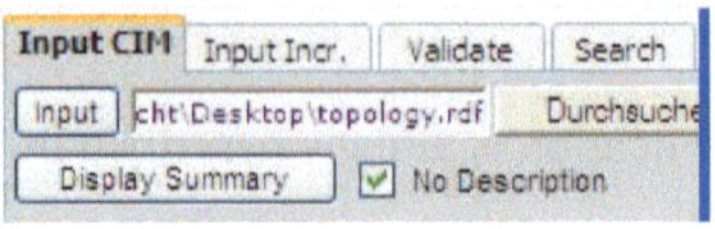

Fig. 6.5 Importing CIM/XML topology

CIMSpy is unsuitable to build a CIM/XML topology of the scratch, so it is necessary to import an existing topology into the tool. After starting up the DHTML page in the browser, the RDF file can be imported can be imported at the "Input CIM" tab on the left site of the page(see figure 6.5.

After importing the tool generates several views as shown in figure 6.6. On the left side a hierarchical view of the imported topology is available. The summary in the main windows presents a table of all elements and the number of objects as well as a description.

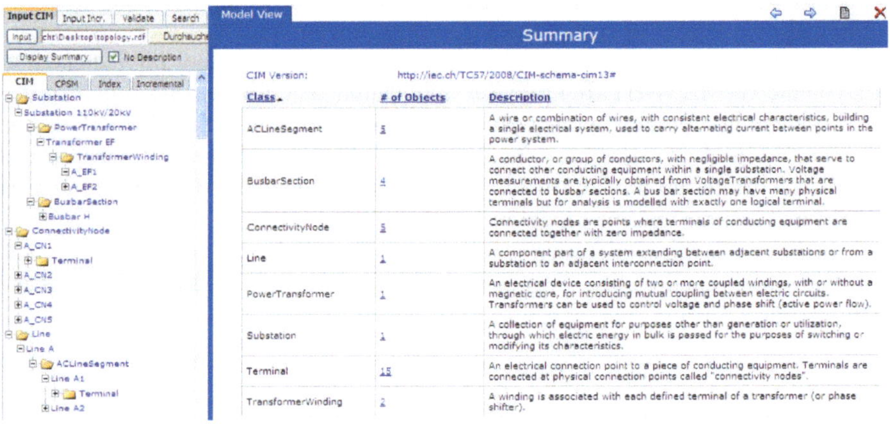

Fig. 6.6 Overview page after importing a topology

The "Input Incr." tab allows to import more than one RDF file in case the topology has been split into several incremental topologies. As the name stats the "Validate" tab allows to validate the imported CIM/XML topology in three different ways. It allows the validation against the CPSM rules, validation against domain modeling rules and the validation against the CPSM OWL rule set. Figure 6.7 shows the result of a validation of the topology generated in chapter 5.2.1.

Is it can be seen the topology produces a few warnings and one error, which with the included description can be fixed if necessary. The last tab provides a mask for searching elements in the model and jump directly in the navigation tree to the found objects.

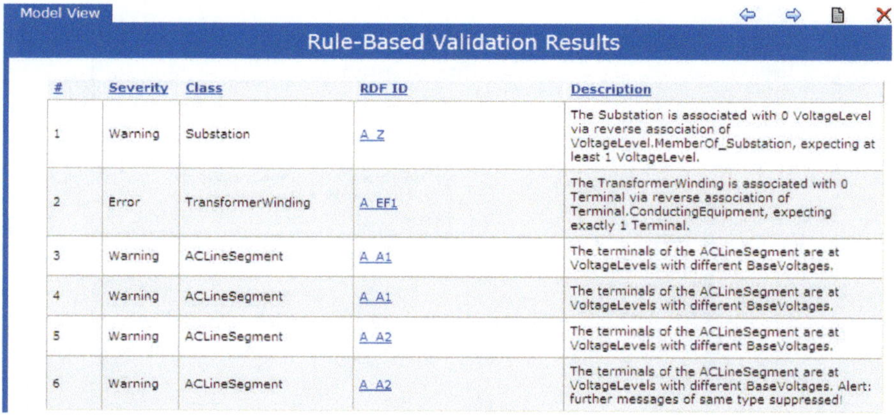
Model View

Rule-Based Validation Results

#	Severity	Class	RDF ID	Description
1	Warning	Substation	A_Z	The Substation is associated with 0 VoltageLevel via reverse association of VoltageLevel.MemberOf_Substation, expecting at least 1 VoltageLevel.
2	Error	TransformerWinding	A_EF1	The TransformerWinding is associated with 0 Terminal via reverse association of Terminal.ConductingEquipment, expecting exactly 1 Terminal.
3	Warning	ACLineSegment	A_A1	The terminals of the ACLineSegment are at VoltageLevels with different BaseVoltages.
4	Warning	ACLineSegment	A_A1	The terminals of the ACLineSegment are at VoltageLevels with different BaseVoltages.
5	Warning	ACLineSegment	A_A2	The terminals of the ACLineSegment are at VoltageLevels with different BaseVoltages.
6	Warning	ACLineSegment	A_A2	The terminals of the ACLineSegment are at VoltageLevels with different BaseVoltages. Alert: further messages of same type suppressed!

Fig. 6.7 Summary of the rule-based validation results

6.2.5 CIM EA

Name	CIM EA
Purpose	CIM modeling and design
Platform	Enterprise Architect Add-in
Short Description	CIM EA extends Enterprise Architect to provide a single environment in which users can manage the IEC CIM, CIM Profiles, and CIM-based artifacts such as RDF and XSD message generation. With CIM EA, users can create and edit CIM Profiles as standard UML models and generate CIM-based artifacts from those models, all within Enterprise Architect.
Link	www.cimea.org

CIM EA is a free Enterprise Architect (see chapter 6.3.1 - Addin developed from Xtensible Solutions, Inc. to provide a single environment for CIM user especially for profile handling without using an external tool like in example CIMTool(see chapter 6.2.3) [130].

6.2.5.1 Features

The main feature of CIM EA is the creation, editing and managing CIM Profiles within the official CIM UML Environment Enterprise Architect. It is possible to create implementation artifacts like RDF and XSD files for further use which are fully IEC 62361-100 (see chapter 1.6.6) compatible. It is also possible to manage different profiles into profile groups directly in EA.

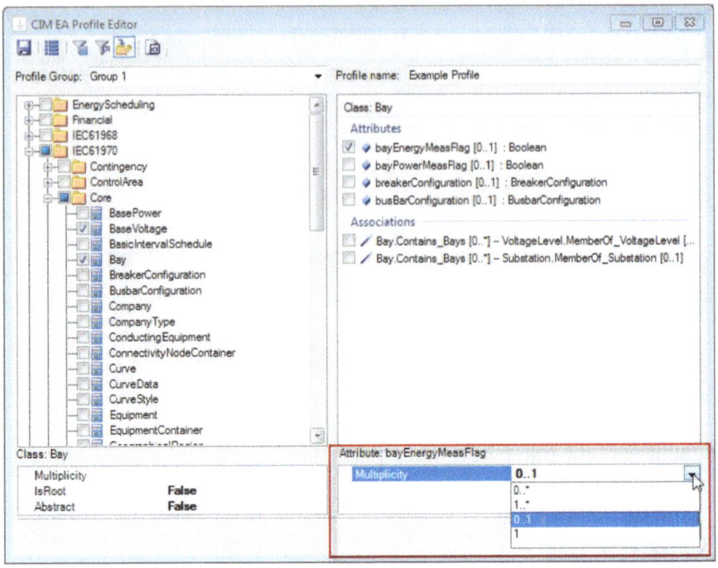

Fig. 6.8 Profile creation and setting of multiplicity[130]

6.2.5.2 Application

After starting up the plugin in EA by using the context menu on the CIM package a new popup-window appears, which holds the UI for CIM EA. This window will provide the necessary options to generate a profile and generating the desired artifacts. Figure 6.8 shows a sample screenshot in the middle of the process of creating a profile. At first, it is necessary to name the profile and select the concerned profile group if needed. At next the wanted CIM classes have to be selected as well as the selection of attributes and associations. Thereupon the multiplicity of the elements as well as the additional options for the root element have to be selected.

After saving the modeled profile, the context menu which appears by a right click on the CIM package, offers the possibility to generate artifacts for the further implementation process. Figure 6.9 shows the artifact generating window. Only a few additional settings are necessary for the creation process like the name or the namespace and of course the selection of the desired artifacts on the left side.

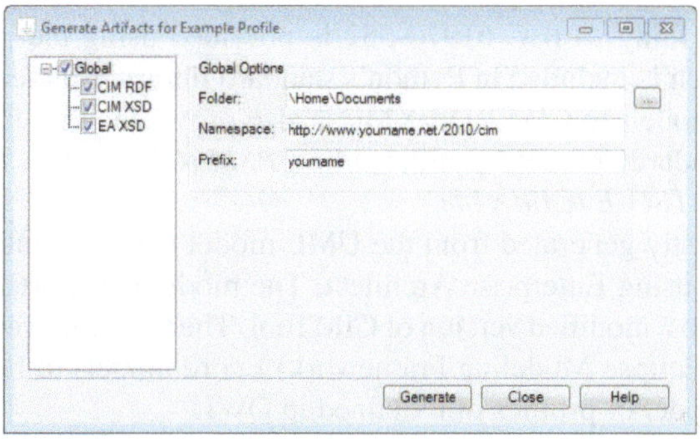

Fig. 6.9 Profile creation and setting of multiplicity [130]

6.2.6 PyCIM

Name	PyCIM
Purpose	Implementation of CIM in the Python programming language
Platform	Python
Short Description	• Support for IEC 61970 15v13 and IEC 61968 11v05 • Legacy support for IEC 61970 14v15 and IEC 61968 10v31 • Profiles of the CIM, including: – Common Power Systems Model (CPSM) (CIM v14) – Common Distribution Power System Model (CDPSM) (CIM v14 and v15) – European Network of Transmission System Operators for Electricity (ENTSO-E) (CIM v14) • Class and attribute documentation integrated as Python docstrings • Transparent bi-directional reference handling using Python properties • CIM RDF/XML parsing and serialization according to IEC 61970-552.
License	MIT
Link	pycim.org

PyCIM is an implementation of the IEC Common Information Model in the Python programming language. It has been developed at the University of Strathclyde, Glasgow by Richard W. Lincoln.

6.2.6.1 Features

PyCIM currently implements IEC 61970 15v13 and IEC 61968 11v05 as well as IEC 61970 14v15 and IEC 61968 10v31. Support for parsing CIM XML/RDF data stored according to IEC 61970-552 is provided using the popular *Element-Tree* library which is included in Python's standard library. Efficient serialization of PyCIM object models to CIM RDF/XML is also provided. The library implements some of the standardized CIM profiles, like *CPSM (CIM v14)*, *CDPSM) (CIMv14 and v15)* and *ENTSO-E (CIM v14)*.

PyCIM is mostly generated from the UML model that is maintained by the IEC working groups using Enterprise Architect. The model is exported in XMI format and imported into a modified version of CIMTool. The CIMTool version is enhanced to allow EMF (Eclipse Modeling Framework) Ecore models to be generated from the full UML model or profiles of it defined in OWL.

Ecore is a metamodel that can be considered a subset of UML designed for code generation. The UML model exported from Enterprise Architect can not be imported into EMF directly due to incompatible interpretations of the XMI standard. However, CIMTool can read the XMI files and creates an internal OWL representation. Transforming the OWL model, the resulting Ecore model is passed to Enerate: a code generation tool based on Java Emitter Templates (JET).

The generated code consists of one Python module for each class in the UML model with all attributes and references associated with the class represented. Due to Python's use of dynamic types and the inability to associate metadata with attributes, private class attributes are used to store attribute types, default values and the names of attributes that are enumerations. This is essential for the PyCIM parser and the serialization routines to function. Bi-directional references are used extensively in the CIM. Python properties are used in PyCIM to provide a transparent method of ensuring that when a reference to an object is set or unset, the opposite reference is updating accordingly.

6.2.6.2 Application

PyCIM has been used in the *GridSurfer* project in conjunction with *Pylon/PY-POWER*[1] and a *cim2pylon* library to perform power flow analyses based on simple CIM topologies. PyCIM was used to read and parse CIM XML/RDF files. The CIM objects were then transformed from the node/breaker to the bus/branch model. The results were passed to Pylon to perform a power flow analysis.

6.3 Commercial Tools

Beside the Open Source Tool several important commercial tools exist supporting the utilization of the CIM. The two most significant ones are introduced in this section. On the one hand, SPARX Enterprise Architect which is used to maintain the CIM and thus plays a key role in the developing process and on the other hand CIMSpy EE/CIMDesk used to enable model exchange between ENTSO-E and its TSO members.

[1] PYPOWER: `http://pypower.org/`

6.3.1 Enterprise Architect

Enterprise Architect (EA) is a Computer Aided Software Engineering (CASE) tool what means that it could be used to design and build software systems, to model business processes and to model any kind of other processes and systems. Therefore, the current version 2.3 of the UML specification is used as basis. UML defines a graphical symbolic language to describe systems during their planning or in cases of existing systems for documentation purposes. EA covers all development cycles and allows a comprehensive traceability of the project from the first draft to the final roll out and even for the following maintenance phase. Furthermore it supports requirements engineering, testing and change management. Figure 6.10 shows the EA representing CIM objects within a class diagram. Since EA is a very powerful tool[2] it could be used for different use cases, e.g., maintaining a large data model like the CIM. Moreover, EA allows the use of add-ins to extend the capabilities of the tool. Developing own add-ins is easy to learn, because a good API is provided by the tool.

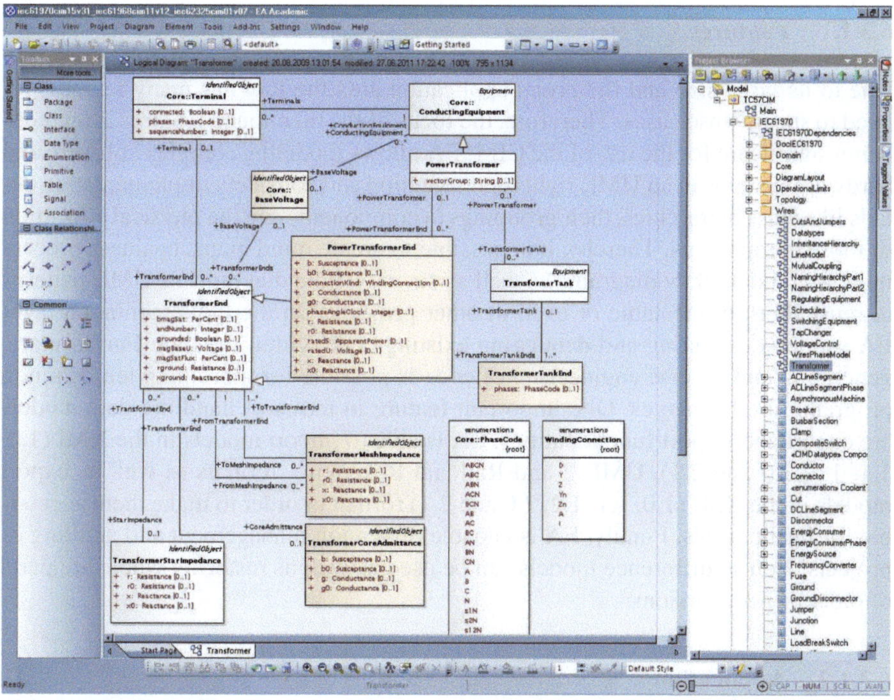

Fig. 6.10 SPARX Enterprise Architect with CIM UML model.

[2] A EA User Guide can be found here: http://www.sparxsystems.com.au/bin/EAUserGuide.pdf

The following table gives an brief overview on EA:

Name	Enterprise Architect
Purpose	UML model management and maintenance
Platform	Dedicated software
Short Description	Visual modeling platform used to manage/maintain the CIM model and design artifacts, such as activity/sequence diagrams, profiles, message definitions. • comprehensive UML analysis and design tool • modeling for business, software, systems • full traceability from requirements to deployment • scalable, team-based repository • enterprise frameworks, Business Process Model and Notation (BPMN), many third party add-ins available
Link	www.sparxsystems.com.au

6.3.1.1 Features

Due to its large size and the broad application area the tool has many features related to specific use cases. Therefore, the focus of the mentioned features is on those being significant for the use of the CIM. EA allows modeling complex software and hardware systems in an UML right representation and considers implementation details like class hierarchies, their groupings to components and the physical allocation of those components. Thereby, it is possible to create mind maps, business process models and data flow diagrams as well as saving whole diagrams as UML patterns to reuse them in the same or even in other projects. On the implementation layer EA supports executing and debugging existing code within the tool. Furthermore, generating and reverse engineering of code is possible for more than ten common programming languages. One important feature in terms of handling data models are export and import functionalities. EA is able to import models in the XMI (1.0, 1.1, 1.2, 2.0 and 2.1), UML 2 and Rational Rose .emx formats as well as export models in the XMI (1.0, 1.1, 1.2, 2.0 and 2.1) format in order to make them accessible for further tools. Finally, EA is capable of version management and auditing of projects. Hence, difference models can be used as well as restoring whole are parts of older project versions.

6.3.1.2 Application

In late 2008, IEC TC 57 WGs 13 and 14 decided to use EA[3] instead of the previously used Rational Rose from IBM. In the following years, SPARX was very active in co-operating with the CIM users group and, thus, with the people being in charge

[3] http://www.sparxsystems.com/press/articles/iec.html

for the development and maintenance of the CIM. They became member of the UCA International Users Group (UCAIug) and attended the CIMug meetings as active participants. CIM related add-ins are strongly supported by SPARX and they consider the needs and requirements of the IEC in a very cooperative way.

EA is mainly used to model and maintain the CIM using UML. Especially class diagrams are used to express the CIM objects as classes with their attributes and being linked with associations. Thereby, aggregations and compounds as well as the generalization concept are used to put the classes into context.

Figure 6.11 shows a very important feature of the EA. Exporting the model as XMI file is the basis for using further tools like CIMTool and CIMBench. Without exporting functionalities it would be possible to realize a further processing of the CIM.

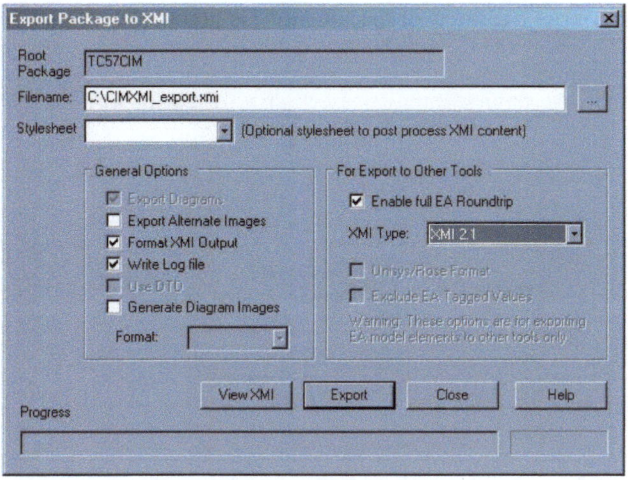

Fig. 6.11 Exporting UML models in the XMI format.

6.3.2 *CIMSpy EE/CIMDesk*

Name	CIMDesk
Purpose	CIM/RDF/XML file browser, editor, and validator
Platform	Eclipse plug-in
Short Description	CIMDesk is a special edition of CIMSpy (see chapter 6.2.4 built for the purpose of ENTSO-e network requirements.
Link	www.powerinfo.us

CIM data engineering solution kit (CIMdesk) is a tool build on the CIMSpy EE tool, which is a tool based on CIMSpy SE with additional features. The tool was especially developed for the European Network of Transmission System Operators for

Electricity (ENTSO-E) to support the CIM profile and the business processes specially designed to support model exchange between ENTSO-E and its TSO members. CIMDesk is a web based tool which is used with the Microsoft Internet Explorer, but contrary to CIMSpy an additional web application server is necessary.

6.3.2.1 Features

Because CIMDesk is build on CIMSpy SE they share the basic features mentioned in chapter 6.2.4. The additional features include most notably the interactive visualization possibilities. CIMDesk is capable of generating single-line diagrams based on the imported data. This feature is rather unique in this domain. It is also possible to modify technical data within the visualization. The validation abilities have been increased compared to the CIMSpy SE which has a limited set of possible error notifications. A brief overview of the functions is graphically shown in figure 6.12.

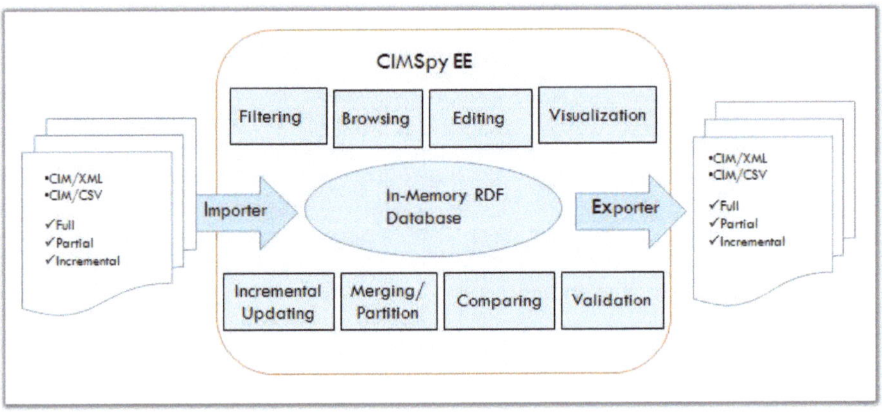

Fig. 6.12 Overview CIMSpy functions from [101]

6.3.2.2 Application

The process of importing the data is exactly the same as in CIM SE described in chapter 6.2.4. Most other functions are also very similar. Because the visualization is a unique feature it will be described further in this chapter.

To use the visualization feature an import of data is needed. The CIMDesk installation comes with sample files for Entso-E specific topologies. In this application example the file "'UCTE10U_ME_r1.xml"'. Figure 6.13 shows a screenshot after importing this file into CIMDesk. The left side shows the hierarchical tree view, already known from CIMSpy SE. To visualize, for example a substation, it is possible to right-click on the specific element in the tree. A context menu appears as shown in the screenshots, which provides the option to visualize the substation in three specific ways. The first one is a simplified view on the selected element, the second one

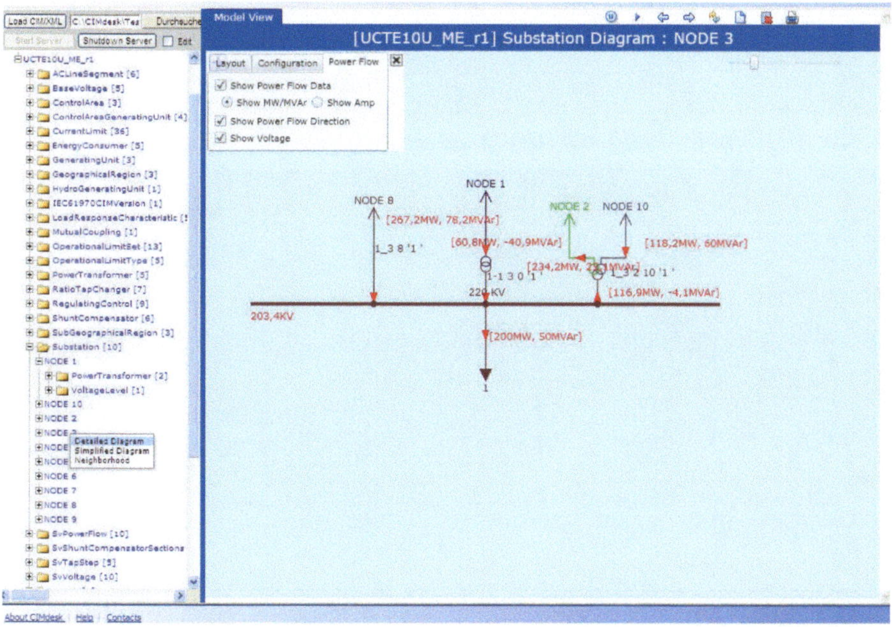

Fig. 6.13 Single-line view in CIMDesk

presents a detailed view and the last one includes direct (up to four level) neighbors of the selected element.

The main screen of the screenshot in figure 6.13 shows the detailed view. On the upper left a additional option menu for the visualization is available. Within this option the view can be adjusted. It is possible to change colors, to show/hide different labels, or even to activate the load-flow analysis which then can be shown as in this screenshot.

Chapter 7
Participation in the CIMug

7.1 Introduction of the CIM Users Group

The CIM Users Group (CIMug) is an association of users of the CIM. Members of the CIMug are utilities, software vendors, manufacturers, consultants and R&D organizations. Currently, over ninety companies, including more than thirty electric utilities, from twenty six countries joined the CIMug. The CIM Users Group was founded in 2005 as subgroup of the UCA International Users Group (UCAIug), a not-for-profit corporation [7]. The CIMug aims at "Connecting The Utility World Through A Common Data Model" [7]. Within this it aims at developing the CIM model consensus and consistency across the industry and hence supporting its members to adapt IEC TC 57 standards [6]. Apart from managing and communicating issues concerning the CIM model the CIMug shares technology basics, best practices and technical resources.

According to [7] the CIMug aims at promoting the CIM, related products and implementations and other IEC TC 57 standards and therefore provides:

- liaison with other standards groups and assess the impact of other standards on the form, scope and content of the CIM
- a central Repository for CIM issues, models, messages, payload designs, schemas and expertise (the CIMug website)
- a single point of contact for CIM Model Management and issue resolution (the CIMug website)
- and a Help Desk for members (the CIMug website)

The central point of information is the CIMug site at `http://cimug.ucaiug.org/`, see Fig. 7.1. The CIMug site is a sharepoint server providing plenty of documents and collaboration functionalities. Here, the latest news on CIM and the CIMug as well as CIM model versions and further documents on standards and CIMug working group information are provided.

General information on the CIM, the CIMug, slides of past CIMug meetings as well as past releases of the CIM are publicly available. Further information and documents requires CIMug membership and login.

M. Uslar et al.: The Common Information Model CIM, POWSYS, pp. 169–177.
springerlink.com

Fig. 7.1 Screenshot CIMug Website

To further promote the use of the CIM, the CIMug organizes two regularly scheduled meetings each year, one in Europe (usually in June) and one in the US (usually in November) [6], here several presentations and events are offered like:

- tutorial sessions like the CIM University providing an introduction to the CIM usually one day before the official meeting
- several presentations on lessons learned and new CIM projects
- status and coordination reports regarding the latest IEC developments like the CIM Model Manager report
- vendor sessions where vendors can demonstrate their CIM-related software or services
- and social events providing a forum for discussions and networking

At CIMug meetings, several sessions and presentations are taking place addressing both people new to CIM, with introductory information like the "CIM University", as well as experts, which are looking for the latest updates and developments on the CIM. The CIMug is organized by experts leading the IEC TC 57 working groups in charge of the development of the CIM. Usually, several presenters and lecturers are IEC TC 57 members, for example the current CIM Model Managers usually attend the CIMug meetings, and therefore reports on the latest CIM and IEC TC 57 developments are provided. Apart from sessions and talks the informal status offer plenty of time to get in contact with prospect customers, experts and users of the CIM to network and share the latest experiences and news.

Registration for a CIMug meeting is done via the CIMug website, see CIMug meeting announcements. Users of the CIM as well as vendors offering Tools for the CIM are invited to give talks and present their projects, tools and company. In addition, non CIMug members are also welcome to take part and may join the CIMug afterwards. For people new to the CIM attending the complete CIMug meeting, especially the "CIM University" or "CIM 101 sessions" are recommended.

7.2 Participation in the CIM Users Group

Companies, public organizations, universities, R&D organizations, associations as well as individuals can join the CIMug, depending on the type of organization, individual or corporate membership and annual revenue different charges exist. Apart from this free guest membership with access to all public information and documents on the UCAIug web-sites is possible. Detailed information on costs and benefits can be accessed at `http://www.ucaiug.org/Pages/join.aspx`. Joining the CIMug is done through a membership of the UCAIug which provides complete access to the the three affiliated user communities CIMug, IEC61850ug and OpenSmartGridug.

As a member the following benefits are provided [7]:

- access to events, forums, task forces, and links to all CIM related resources through the CIMug website
- support for product testing certification activities to lower implementation costs
- access to information on standards activities, educational materials, technical documents and best practices from existing users
- the possibility to provide input to international standards activities at minimal cost and effort
- members-only restricted access to technical issue resolutions, industry experts via a global help desk and on-line discussion forums
- valuable insight from peers and benefit from sharing common solutions
- discounts on paid seminars and exhibitions
- Support and participate in a community effort to establish useful standards for our industry
- marketing and promotion by making company's leadership visible in the industry

As a CIMug member plenty of information is accessible and even not being directly involved in IEC standards development draft IEC Standards related to the CIM like IEC 61968 or 61970 are accessible as well. In addition CIM experts can be contacted directly and through exchange of experience or problems the development of the CIM is influenced without having to attend international standardization meetings.

CIMug members can post CIM issues regarding the electronic model directly through the CIMug website. Here, a special form is provided where errors, problems and proposed changes can be posted, see Fig. 7.2.

This form requires a detailed description of the problem, here providing an example and use case is recommended, as well as a proposed solution. Once an issue is submitted its resolution is done according to the CIM electronic model life cycle as illustrated in Fig. 7.3.

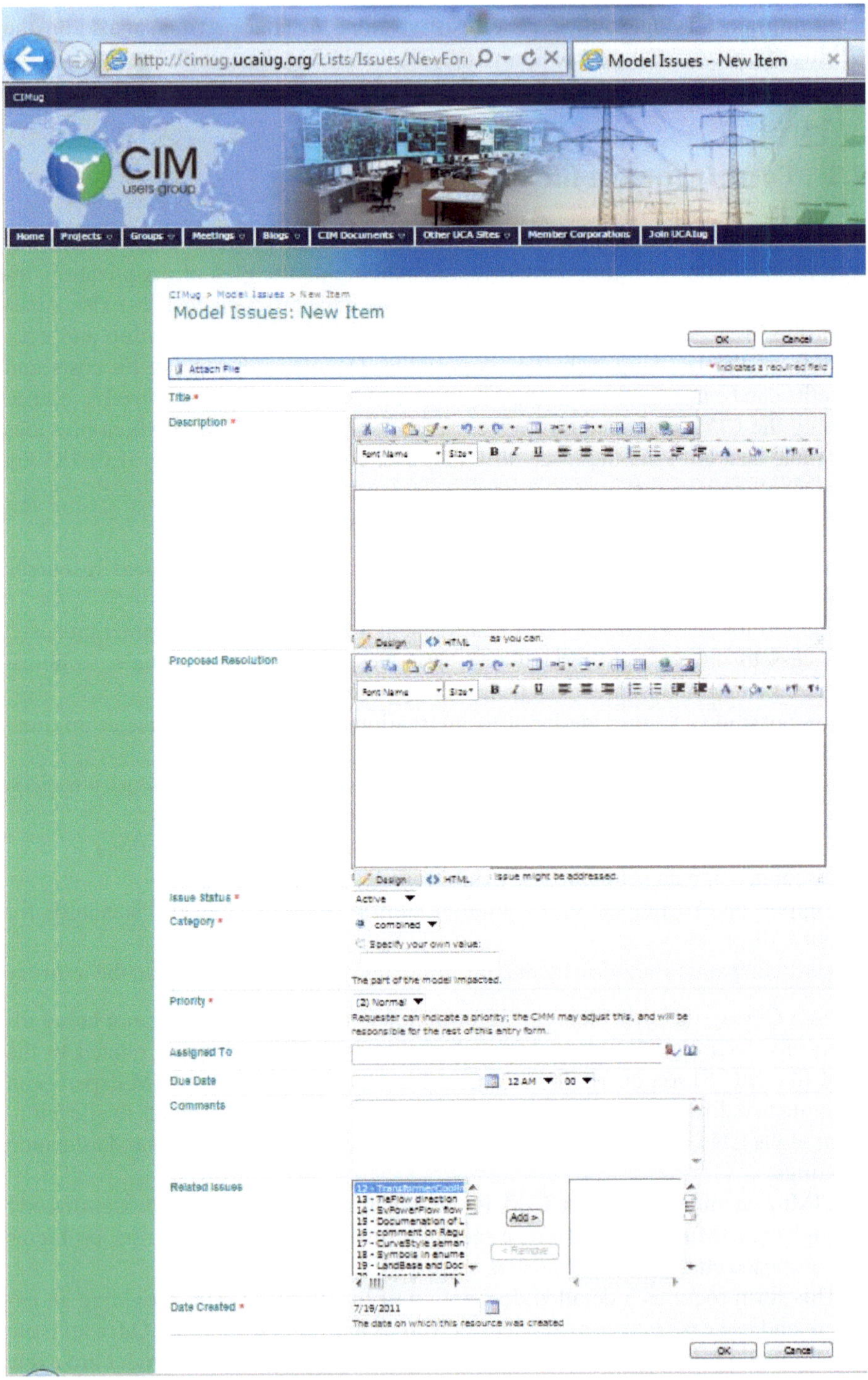

Fig. 7.2 CIM issues form

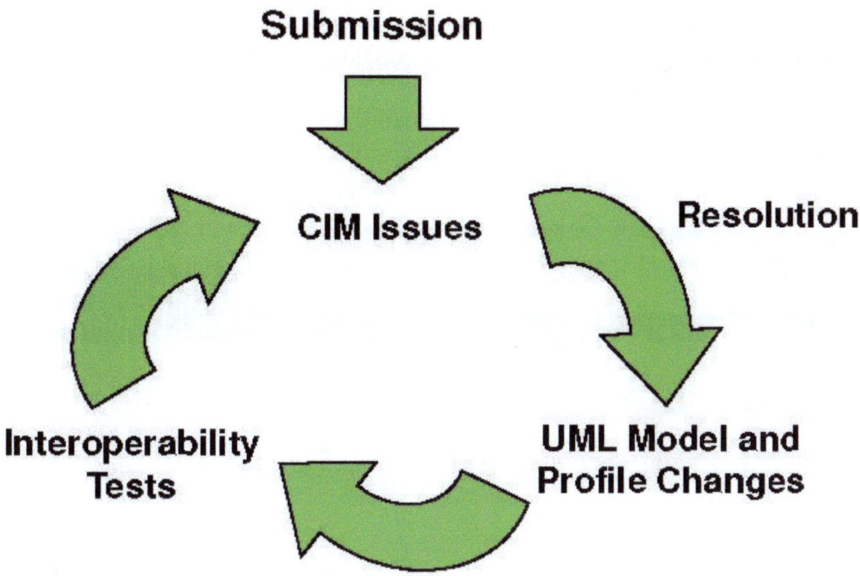

Fig. 7.3 CIM electronic model lifecycle [98]

After submission of an CIM issue the issue is added to the current CIM issues list where it is given a priority. Depending on the complexity and priority of the issue its resolution will be addressed sooner or later and discussed by CIM modeling experts, mainly by members of the CIM Modeling Team, within CIM Modeling Team conference calls and meetings. Here either proposed changes are accepted amended or declined and the submitter is informed. CIM issues might lead to model or profile changes which will be published within a new CIM release, see also section 1.8.5. New releases of the CIM will then be used within interoperability tests and might lead to new issues.

7.3 Participation in National and International Standardization

Apart from CIMug participation joining national and international Standardization is also possible. Taking part in international standardization at IEC TC 57 WGs requires involvement within the corresponding national standardization organization or NC. For example in Germany, the NC responsible for the IEC TC 57 working groups is the DKE. Within the responsible NC several working groups might exist which requires to get in contact with the corresponding working group (for Germany the K 952). Participation within national standardization is voluntary but requires at least attending meetings, discussions and voting at national level. Active contribution can be done through commenting as well taking part in the development of national and international standards. As several groups exits addressing different topics membership in several working groups might be necessary. On the other hand the latest updates and news are available and usually access to draft standards and other related publications is provided. In addition, experts within the working groups can be directly contacted.

For example, in Germany, access to documents is provided through the DKE website, see Fig. 7.4.

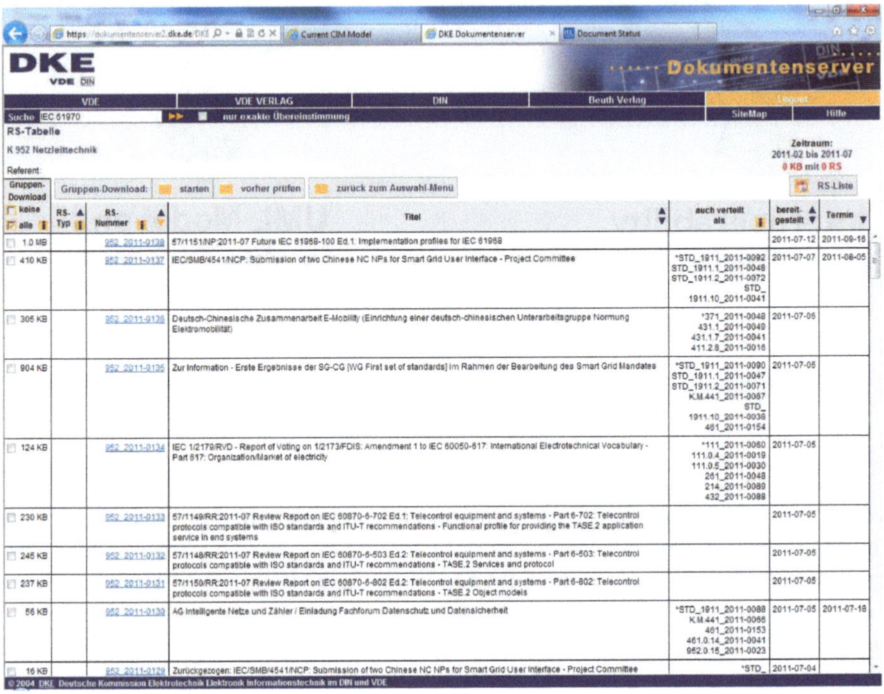

Fig. 7.4 Screenshot DKE Website

Here the latest documents can be accessed and searched for. In addition new documents and meetings are announcement via a mailing list. For taking part in international standardization nomination as WG member by the NC is necessary. This is done in a voting process by the corresponding working group members. When accepted access to the IEC TC57 WG website is provided, see Fig. 7.5.

Here, again the latest document and information on international level can be accessed. Regarding new documents subscribing of alerts is possible. Taking part in international standardization requires the attending of meetings and conference calls at international level as well as the attendance of national meetings to report about current international activities. On the other hand, international standardization can be directly influenced and international experts contacted.

Taking actively part in standardization involves not only attending national and international meetings or (online) conference calls but also commenting, reporting,

Fig. 7.5 Screenshot TC57 WG14 Website

discussing and developing standards related documents. Latest news and discussions are provided through regularly mailings and a lot of communication is done via direct mailing and telephone contact between working group members. Physical meetings are used to present and discuss as wells as to promote current developments face to face and speed up the development of standards. Even though participation involves a lot of work valuable insights, direct contact to other experts and latest information is gained.

7.4 Interoperability Tests

The aim of UCAIug IOP Tests is to demonstrate that different vendor systems are able to successfully exchange models using the IEC Interface Standards 61970 and 61968. As described in section 1.8.5 the aim of both standard series is to integrate applications in the energy industry using the CIM as common data model. IOP Tests use specific CIM profiles (see section 3.8) to perform compliance and conformity tests.

In the past, the testing has mainly focused on three areas:

1. exchanging power system network models for transmission (CPSM, ENTSO-E) and distribution (CDPSM) using the CIM

2. compliance and interoperability testing of the GID standards
3. exchanging messages based on the IEC 61968 standard parts (mainly metering 61968-9)

Figure 7.6 shows the general IOP test procedure.

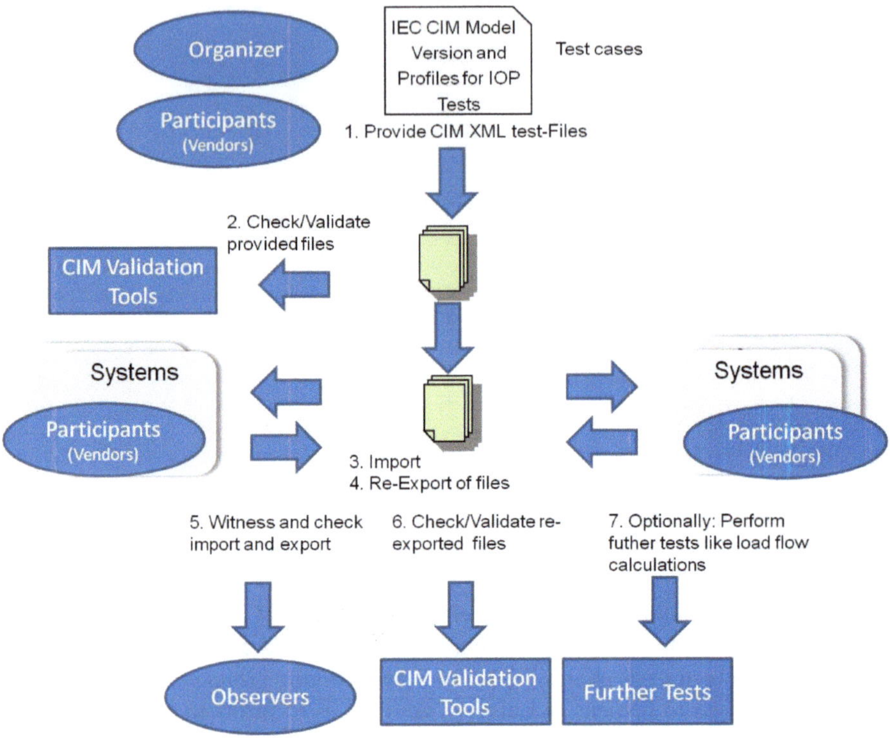

Fig. 7.6 IOP Tests general overview

IOP Tests are organized by IEC TC 57 WG and CIMug according to industry needs and usually rely on the latest agreed combined Version of the CIM. Therefore, IOP Test are conducted after CIM model freeze in Q1. During IEC TC 57 WG meetings dates and locations for IOP as well as test cases and CIM profiles to be tested are agreed on. Based on this through IEC WG and CIMug mailings WG and CIMug members are invited to take part in IOP Tests. CIMug or IEC TC 57 WG membership is expected by participants non-members might take part as well.

Mainly three types of actors can be distinguished within IOP Tests:

- participants: vendors who want to test their applications
- observers: witnessing the test

- validation experts: supporting the use of validation tools like CIMSpy or CIMDesk
- organizer: hosting and organizing the meeting as well as preparing the report

At the moment no standardized procedure model for conducting IOP-Tests is available. In the following basic steps within interoperability tests based on several descriptions of IOP-Tests are provided.[1] Individual IOP-Test my vary and might change in the future as the procedures are always agreed on by the organizing WG members and the participants.

As a prerequisite within WG meetings the latest CIM version, test cases and CIM Profiles which should be subject of the IOP Test have to be defined and organizational aspects (like host, dates, location, participants and report publication) agreed on. Then CIM XML test files have to be provided by each participant (Step 1). All provided files need to be validated against the underlying CIM profile and corrected in case of violation (Step 2). CIM validation tools like CIMSpy or CIMDesk are usually used. On basis of the validated files import and re-export (Step 3 and 4) are performed by the applications of the participants. Observers act here as witness to check the import and re-export. Re-exported files will usually be validated again with CIM validation tools. Participants may decide to only take part in dedicated test cases. Regarding on the type of IOP tests and data exchanged further tests might be conducted. For example when dealing with the exchange of network models load flow calculations might be performed and compared.

[1] See for example EPRI and ENTSO-E IOP test descriptions `http://mydocs.epri.com/docs/public/000000000001013295.pdf`, `http://mydocs.epri.com/docs/public/000000000001012494.pdf` or `https://www.entsoe.eu/fileadmin/user_upload/_library/news/CIM_IOPs_and_Roadmap_Explanatory_note.pdf`

Chapter 8
Perspective

8.1 Introduction

Within this chapter, we provide an insight to the new OPC UA Unified Architecture and a mapping of the CIM to its Address Space Model which was developed using a Sparx EA Enterprise Architect plug-in. To assess the proper use of the CIM, the XML Naming and Definition rules have to be obeyed as well as the profiling mechanisms established by the model maintainers. Testing and conformance checking with the CIM are therefore issues coming up. We will introduce ideas very similar to those of the IEC 61850 IEC TC57 WG 10 to deal with those issues when working with the CIM. Another aspect which is quite related to the implementation of the CIM being an enterprise-wide used data model (domain ontology) at run-time level at the message-bus used, there is a need for implementing security strategies for the CIM using the IEC 62351 standards series and other ontology-related mechanisms. Finally, we conclude the outlook with an overview on future applications and possible extensions of the CIM.

8.2 The OPC Unified Architecture

For the CIM, different communication mapping exist. The previous section of this book have outlined the current application of the IEC 61968 Interface Reference Model using XML and the topology data exchange using W3C's RDF Resource Description Framework. Figure 3.6 already depicted the relationship between CIM and Classic OPC standards. Also in section 3.6.4 a future standard was introduced dealing with a mapping between CIM and OPC UA. Due to its promising application in Smart Grids, this chapter describes the approach of CIM-based OPC UA communication more detailed as it was introduced in [104].

The OPC UA specifications are developed by the OPC Foundation[1] and partly standardized by the IEC 62541 [55]. The so called "Classic OPC" - the predecessor of OPC UA - is well accepted and applied in industrial automation. There are over

[1] http://www.opcfoundation.org/

M. Uslar et al.: The Common Information Model CIM, POWSYS, pp. 179–186.
springerlink.com

22.000 OPC products on the market from more than 3.500 companies, including all major automation vendors [4]. As a consequence, almost every system targeting industrial automation implements Classic OPC.

OPC Unified Architecture unifies the existing Classic OPC as depicted in figure 8.1 and brings them to state-of-the-art technology using SOA. Two of the main advantages of the OPC UA are the platform independence and the capability to use web services. Thus, OPC UA can be run directly on controllers seamlessly integrated into Manufacturing Execution Systems (MES) and ERP systems running on Unix / Linux using Java applications.

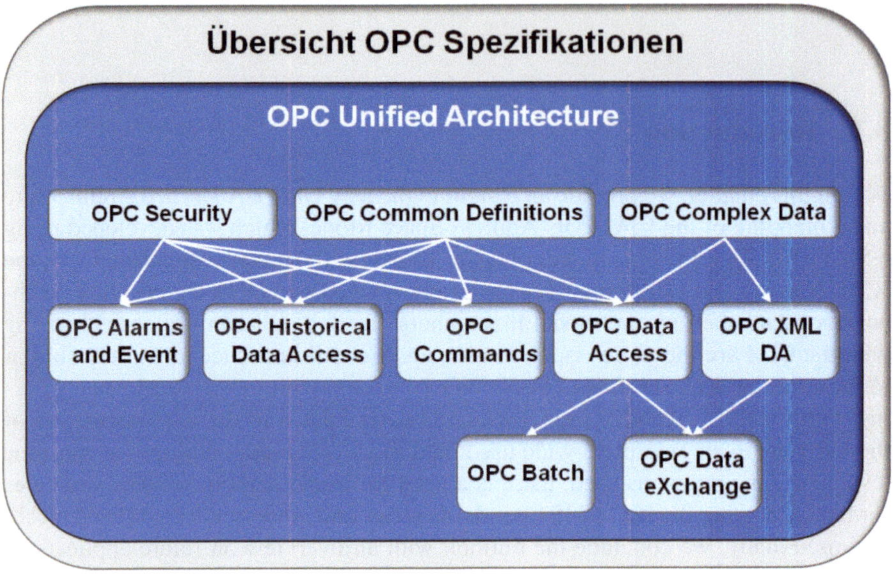

Fig. 8.1 Overview on standards from the OPC Foundation (based on [86])

The object- and service-oriented paradigm as well as the generic information model allows OPC UA to be applied outside of the borders of industrial automation. The new information modeling capabilities enable OPC UA communication independent from a specific domain. Standardized domain specific domain models can be used to be set on top of the generic OPC UA information model. The base principals of OPC UA information modeling are [87]:

- **Using object-oriented techniques including type hierarchies and inheritance.** Typed instances allow clients to handle all instances of the same type in the same way. Type hierarchies allow clients to work with base types and to ignore more specialized information.
- **Type information is exposed and can be accessed the same way as instances.** The type information is provided by the OPC UA server and can be accessed with the same mechanisms used to access instances.

- **Full meshed network of nodes allowing information to be connected in various ways.** OPC UA allows supporting various hierarchies exposing different semantics and references between nodes of those hierarchies. The same information can be exposed in different ways, providing different ways to organize the same information depending on the use case.
- **Extensibility regarding the type hierarchies as well as the types of references between nodes.** OPC UA is extensible in several ways regarding the modeling of information. Besides the definition of subtypes it allows - for example - to specify additional types of references between nodes and methods extending the functionality of OPC UA.
- **No limitation on how to model information in order to allow an appropriate model for the provided data.** OPC UA servers targeting a system that already contains a rich information model can expose that model "natively" in OPC UA instead of mapping the model to a different model.
- **OPC UA information modeling is always done on the server-side.** OPC UA information models always exist on OPC UA servers, not on the client-side. They can be accessed and modified from OPC UA clients. An OPC UA client is not required to have an integrated OPC UA information model and it does not have to provide such information to an OPC UA server.

Figure 8.2 shows how the CIM could be used as domain specific information model for the energy domain.

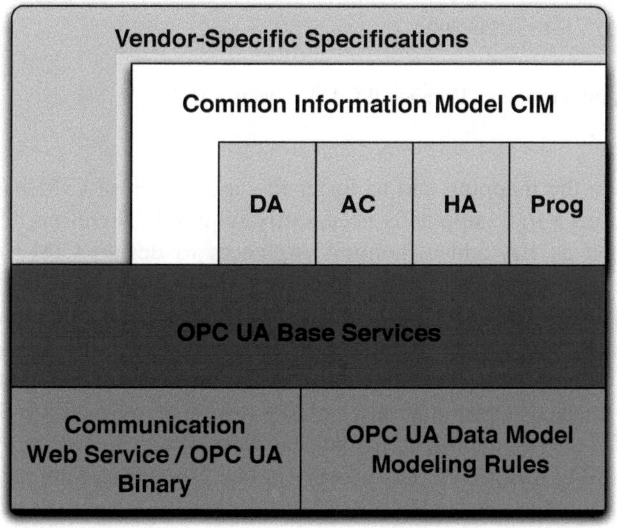

Fig. 8.2 CIM integration into the OPC UA communication architecture (based on [87])

To realize a CIM-based OPC UA communication architecture it is necessary to map the CIM-objects to the OPC UA elements as shown in figure 8.3. Therefore, the following mapping is suggested, based on the content of [53]:

- CIM classes will be mapped to OPC *ObjectTypes*. Ineritance of CIM classes are expressed by OPC *HasSubType* references within the related *ObjectType*.
- CIM attributes can be mapped to both variables or properties (depending on the IEC mapping rules and design decisions).
- CIM associations will be mapped to OPC *ReferenceTypes* which will be referenced to the associated *ObjectTypes*.
- CIM data types (enumeration, compound, etc.) will be mapped to OPC *DataTypes*. However, primitive CIM data types (*String*, *Boolean* etc.) don't have to be mapped explicitly because existing OPC *BuildIn-DataTypes* can be used.

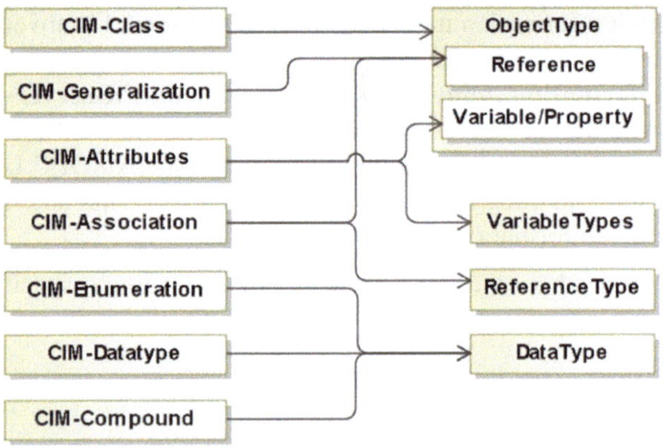

Fig. 8.3 Mapping between CIM and OPC UA objects [102]

To facilitate the mapping and to foster the application of CIM-based OPC UA communications a tool support is necessarily required. Therefore, CIMbaT [103] was developed as EA add-in helping engineers to design CIM-based Address Spaces. The main component of the tool is a CIM-to-UA Address Space mapping which is constructed as a step-by-step wizard. The mapped elements will be stored into a XML-file. The wizard includes a XML configuration file in which the main and default settings can be done manually but also be manipulated in the first wizard step. During the next steps the wizard includes design-steps for editing the mappings for single CIM classes and their attributes as well as associations in which the user got the possibility to set OPC UA properties like *IsAbstract*, *SupportsEvents*, *Historizing* and *DataType*. It is also possible to decide for a designer whether a CIM class attribute should be mapped to a OPC *Property* or *DataVariable*.

The manipulable CIM elements can be easily selected within a tree-structure and the editable values are displayed on the right-hand side (see figure 8.4). Each added OPC UA specific information will be realized as a unique stereotype of the according CIM object. Because the additional information will be saved as UA-specific stereotypes in the CIM model they will be recognized when restarting the wizard thus the design choices are preserved. Now, a generic Address Space could be generated or the wizard could be used to create instance of the abstract objects to develop a completely instantiated model for an OPC UA server.

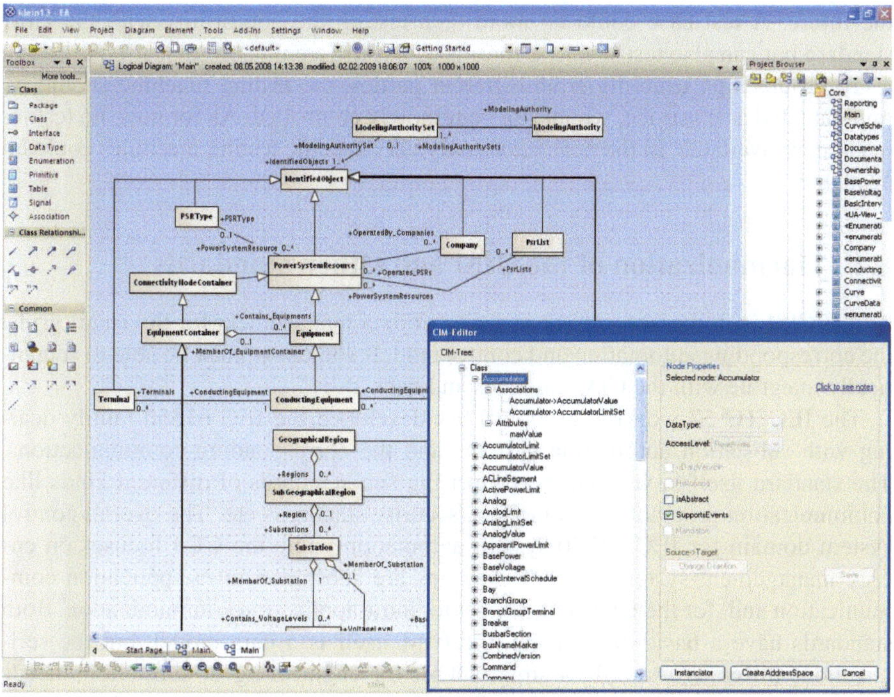

Fig. 8.4 Design decision within CIMbaT while creating an OPC UA Address Space

8.3 Testing CIM Payloads

The CIM and the included interface specifications are a very complex system in its own. Standards and specifications are often formed in natural language, which leads to more than one interpretation of the content. Another point is that standards can have gaps, errors or even contradictions which can not be found until the implementation phase. A specific problem can be solved in different ways, so that a standard (data model) may be not enough to ensure a harmonized communication.

The interoperability between different systems from different vendors or different implementations only is even more complex. To ensure the interoperability the aforementioned IOP tests in chapter 7.4 are a first step but are expensive, protracted, and needs the participation of all affected players. The increasing number of participants and interfaces lead to an increase of expense through an higher number of tests necessary.

One way to minimize the effort needed could be the implementation of a testing machine. Automatically performed tests can be repeated as often as needed from different parties, like vendors, developers and users equally.

A testing machine should be able to test a wide range of interfaces based on CIM. The system itself should be expandable to ensure further technical developments in the future. A test case could be more then testing the conformity with the CIM standard but can also test whole scenarios across different communications.

To ensure high usability from different parties, the testing machine should be implemented as a service, which can be remotely controlled. As for now, no testing machine is available. In the near future only very specific testing machines could be available.

8.4 Harmonization of the CIM and Other Standards

As the CIM is in the very center in terms of data management for the smart grids, the corresponding automation and control models somehow must be able to semantically integrate with the CIM payloads and data.

The IEC TC 57 working group 10 has developed the IEC 61850 family dealing with substation automation systems and the corresponding communications. The standard itself is very large, comprising sub-standards of different kinds like communication protocols, data models, security standards etc. The overall control system domain for IEC 61850 system automation while the CIM focuses on energy management systems, the IT domains are substation intra-application communication and, for the CIM, control-center intra-application communication. Both standards have a basic data model, the CIM itself is a data model, but the serializations differ. While only a small subset for engineering of substations needs a XML-serialization in IEC 61850, all CIM objects can either be serialized using XML, RDF or OWL. Also, the IEC 61850 family lacks a significant support of APIs. Those differences lead to problems when coping in real-world projects with the IEC TC 57 reference framework.

The data from both standards must be used in the same context, therefore, a seamless integration of i.e. the functional description structure of a substation must be known to the SCADA and a mapping between structure from IEC 61850 and IEC 61970 is needed. Furthermore, all data points and measurements from the field devices must be mapped from IEC 61850 semantics to IEC 61970 semantics. Without any doubt, those two scenarios are the most striking ones, but further scenarios exist.

Unfortunately, problems occur when trying to use the standards. The different WGs have used different naming schemes, object-oriented modeling vs. hierarchical modeling, different semantics, different tools and serializations and, the IEC 61850 model does not exist as an electronic model as of today (although an initiative on bringing 61850 to a UML model in Sparx EA has been started), just within tables in a Word document. Finally, all the standards have been made final international standards; therefore they are being implemented by vendors who rely on the stability of the standards and their products. Their existing implementations cannot be harmonized at the meta-model level breaking several aspects for a new, overall harmonized standard family comprising IEC 61850 and IEC 61970. Users have to deal with harmonization on a conceptual mediator layer.

Since 2000, different approaches and reports have dealt with the problem of integrating 61850 and CIM. Starting with unified models which were suggested by EPRI reports like [21], [15], [17] or industrial work from ABB [85], [84], viewpoints on how to cope with the issues differed. [113] provided an overview on this very topic and outlined the different scopes which should be addressed with regard to use cases. Recent work alongside this basic overview by [106] has extended the scope and led to a thorough integration of SCL and CIM topology models using formal semantics and reasoning.

Apart from the IEC 61850 integration, standards and models like DLMS/COSEM, MultiSpeak, UN/CEFACT CCTS and Electronic data interchange EDI messages are candidates to be harmonized with the CIM. From the IEC TC 57 perspective, this work is done by the WG 19.

8.5 Envisioned Extensions

As mentioned already in chapter 1.8 the CIM standard is constantly under development and will be extended regularly. In the near future, there are several upcoming extensions like the following:

- **CIM for Weather:**
  At the CIMug meeting in Prague 2011, the extension model CIM for Weather was introduced. The intention is to extend the CIM to cover weather information. The pervasive use of weather information in utilities make this extension to the next logical step. There is a need to exchange weather data (metered, historical or forecasts) to allow accurate load forecasts, to plan deployments and to generate generate forecasts especially for DER. Different existing data models are investigated like Typical Meteorological Year (TMY3) and Weather Exchange Model (WXXM) as well as previous work from industrial partners. These will be used as input to form a data model extension which will be modeled using CIM modeling conventions and existing CIM classes.
- **CIM for Distributed Energy Resources:**
  The integration of DER into the power grid will be on the rise in the next years. Therefore an integration of this generating units in the ICT infrastructure is inevitable. CIM for DER is based on the automation standard IEC 61850-7-420 and will reuse many of the existing objects from there. Just like in the previous described CIM for Weather the model will be generated by using already existing CIM modeling conventions and CIM classes.
- **CIM for Plug-in Electric Vehicles:**
  Just like the DER the PEV will also be more and more present in the near future. The same arguments as used in previous CIM for DER paragraph for modeling an extension can be used here. The PEV extension will not be based on an existing standard but will also use existing CIM modeling conventions and CIM classes.

8.6 Home Automation and Metering

One relatively new application area is the HAN. For a seamless Smart Grid ICT infrastructure, the home automation is at the bottom line of the communication path and has to be integrated. Different approaches are taken by different parties and at the moment there is not a clear favorite. Only one thing seems to be clear. A communication with CIM systems has to be established in the communication path, so it is necessary to have communication mappings from in house automation to CIM. This would require at least harmonization between the CIM and the used HAN standard.

To minimize harmonization efforts, standards like the ZigBee Smart Energy Profile (SEP) 2.0 have been created. Many items used in the data model are already existing in the CIM data model. Model and design conventions from the CIM domain have been used to. So the harmonization would be less of a Problem. Existing AMI standards like the DLMS/COSEM data model are currently under harmonization work and results for a joint TC 13 and TC 57 effort can be expected in the next time.

Appendix A
Use Case Modeling for Smart Grids According to IEC/PAS 62559

Introduction to Use Case Modeling for Smart Grids

This chapter is based on both a project and a report the authors have prepared for the German DKE. The following aspects and requirements have been discussed and developed in German national standardization and are envisioned to be brought into the M490 mandate of the EC and the IEC SG 3 work.

Use cases were originally established for the purpose of identifying how a system behaves in relation to stakeholders. For CIM-based communications, the establishment of use cases is to document future scenarios of energy management systems. A uniform and structured documentation also provides a basis for communication between actors involved in system development, and the analysis of the use cases permits deduction of requirements for system development. Starting with the requirements defined, the conceptual design of data models, interfaces, exchange processes and protocols can take place afterwards. As the result of these efforts, the artifacts identified are to be studied with regard to their need for standardization and involvement in the standardization process.

With a sound base of use cases which cover large portions of the future energy systems that are relevant to standardization and with the resulting requirements, a basis for the design of interoperable technical solutions is created. Standardizing these solution concepts, finally, contributes to the establishment of a standards-based, open and interoperable architecture for a future energy system in which solutions from various manufacturers jointly implement the vision of a smart grid.

Alongside general use cases relevant to standardization as are described here, detailed use cases specific to individual companies and relevant to competition are necessary within those companies to enable them to develop solutions and sell them on the market. The recommendations presented in this annex are also, with some limitations, applicable to the description and management of company-specific use cases. The objective must be to support linking the two types of use cases in the long term, in order to enable companies to analyze their use cases in relation to relevant standards and, thus, promote standardization work and the development of interoperable technical solutions.

This appendix initially describes requirements for the compilation of use cases and presents recommendations in that regard. Then, on the basis of IEC/PAS 62559, a strategy for use case compilation is described and a template for use case compilation is outlined. For careful description and handling of various use cases, a classification system and conventions intended to support the process of use case identification are presented, followed by a five stage process for compilation and processing of use cases. In addition, the embedding of use cases in international standardization is outlined.

Requirements for Use Case Descriptions for the Smart Grid

Due to the complexity of the smart grid, and especially because of the large variety of both actors and systems, use case descriptions are to be used to create a common understanding for the development and identification of interfaces and data models. In this context, use case descriptions are to be established for various target groups, such as

- companies in the energy industry,
- IT manufacturers,
- equipment manufacturers,
- standardization organizations,
- legislators, and
- companies from other sectors than the energy industry.

This implies that use case descriptions for such an extensive group of stakeholders can only be established involving several participants with different background knowledge and viewpoints. Use case descriptions represent the central element for the description of requirements and the general functionality of systems, and they are to support cooperation between various experts (e.g. electrical engineers and IT experts), sectors of industry, organizations and committees. Furthermore, use cases form the basis of further standardization work and the design of interoperability and conformance tests. As a result of the complexity involved, clear methodology and classification and corresponding support by software tools and tool chains is required. National use case descriptions are also to be used to compile corresponding profiles for standards with the aim of refining international standards for use, for example, in Germany in relation to German regulatory requirements. In this context, the various levels of national and international standardization and the associated documents are to be observed.

In order to facilitate management of a large number of use case descriptions, possibly compiled by different persons with highly differing backgrounds, a structured, uniform description is necessary if consistent and structurally similar descriptions are to be created. A high level of quality and the opportunity to locate and retrieve use cases are required if the effort involved in their documentation is to be justified.

To provide for contributions to international standardization and to avoid additional consequential costs, it is advisable to present and discuss methods and model procedures for the compilation of use case descriptions in the various standardization organizations on the international level and canvass for their acceptance. This is address e.g. by the Use Cases and Sustainable processes team for the SG-CG M/490 mandate in Europe. User-friendliness of the methods is required, as are reasonable time and costs for compilation of the descriptions, especially as far as the compiler's time and the costs of the tools (TCO) are concerned.

From a technical point of view, the compilation of use cases is to be regarded as a collaborative task. In this respect, functionalities for multiple user operation and systems for the locking, commenting, release, revision tracking and version management of use cases are to be provided. In addition, further aspects such as multilingual capability could be of interest for using use cases and systems in international standardization and projects.

The large number of use cases to be expected gives rise to the following requirements:

- Organization systems: Provision of functionalities for classification, grouping, searching and navigation of use case descriptions.
- Consistency: In this regard, the objectives are to ensure uniform semantics and avoid redundancies. Both of these could, for example, be achieved by using a comprehensive model. It is also to be ensured that the descriptions are complete.

Ideally, the entire process of use case compilation should be supported from start to finish by a single integrated tool. As this usually can only be implemented at considerable cost, import and export from and to other applications should be supported.

General Recommendations on Use Case Compilation and Management

Compiling, updating and managing use cases is complex, especially with a large number of use cases. In this context, a suitable scope and the degree of detail are first to be specified and the descriptions limited. With regard to the requirements described in the previous section, the time and cost of compilation are to be minimized and the process is to be supported by software. It must be taken into account that, apart from the initial compilation of use cases, further use cases will have to be added and existing use cases adjusted. With regard to the maintenance of the use case collection, suitable criteria for classification have to be established. Attention has to be paid here to ensuring a relatively long-term applicability (resistance to change) of the criteria, so as to avoid regular reclassification.

In order to master the challenges associated with the compilation and management of use cases, it is recommended that the process be structured by implementation of the following measures:

- Use of conventions: Guidelines, for example, of naming conventions and particular UML diagrams for the presentation of graphic representations of the real-world.

- Limitation to the essentials: Avoidance of redundancies, for example, and reduction to relevant information only.
- Use of familiar/proven methods/approaches: E.g. relying on familiar/proven modeling languages such as UML.
- Tool support: Use of software tools to reduce manual work, especially for quality and consistency assurance.

In addition, the following procedure is proposed for the establishment of a strategy for use case compilation. It is strongly advisable to take previous related approaches, such as the following, into account.

- German Standardization Roadmap [12] in particular with its classification of standards.
- IEC/PAS 62559 IntelliGrid Methodology for Developing Requirements for Energy Systems [54]. This publication supplies a methodology with corresponding templates for the identification of requirements and description of use cases, and is to be adapted and refined with regard to national needs.
- Established approaches from academic research and practice: In this context, artifacts, methods and procedures are to be examined, for example reference models from the power sector, the catalogue of reference models for the energy industry (Energy-RMC) [22] (with regard to structuring), NIST [92] (domains), T&D Europe (overview) [16], Smart Grid.

Additionally, there is the opportunity for an exchange of information with other (standardization) organizations which are also dealing with the compilation of use cases for smart grids, for example to draw upon the work of the FG Smart group at ITU-T (International Telecommunication Union Standardization Sector).

Furthermore, to order and facilitate retrieval of the use case descriptions, a classification of use cases should be established. See the section entitled Organization and classification of use cases. In order to ensure a unified description of use cases by several editors, it is proposed that a template for use case writing be established, and several templates may possibly be required for different types of editors (e.g. managers, energy experts, IT experts, standardization experts or legislators). In general, the envisaged process of use case compilation and application has to be taken into account, as this should have corresponding effects on the nature of the classification criteria and templates.

The concepts, templates and classification criteria then established will certainly require iterative, continuous elaboration and adjustment. The adjustments required will, in the authors' experience, predominantly become obvious in the course of actual, practical use. Provision is to be made in advance for revisions to these central documents: These should preferably take place in the background, so as not to impede the use case compilers.

Strategy for Use Case Compilation

A number of conventions and guidelines are to be considered, especially if the management and use of various use cases from different authors and sources are to take place consistently. A conceptual description of central components in use case compilation is presented below.

The description of use cases is fundamentally to take place on the basis of IEC/-PAS 62559 [54]. In addition, the use of a central repository for a general template and various role-specific templates is recommended - for both enterprise and standardization use. The additions to IEC/PAS 62559 [54] are described in brief below:

- Central repository: Here, at a central location, various descriptions of, for instance, definitions, actors, conventions and relevant documents are to be stored. These elements are to be accessed and used accordingly during the compilation of use case descriptions.
- General template for use cases: This constitutes a complex, internal template which contains all the relevant attributes for all roles involved. For simplicity s sake, it is advisable to confront only use case experts and equivalent administrators with the general template. A role-specific template should be provided for normal use case compilers (see next point).
- Role-specific templates for use cases: Due to the complexity of use case compilation, specific templates should be established for each role and made available specifically to the various roles such as managers, energy experts, IT experts or editors/use case managers. Clarity for the use case compilers is ensured by the use of role-specific templates.

The structure of the central repository and that of the general template are roughly described below and the additions to IEC/PAS 62559 are presented.

Central Repository

The structure of the central repository is presented below, with the (German DKE) additions to the existing IEC/PAS 62559 accented in bold type.

- **Global Repository**
  - **Glossary**
    - **Terms, List of Roles (market roles / system roles), Actors List, Acronyms**
  - **Information Exchange**
    - **Data Objects, Data Protection Class,** Characteristics of Information Exchanges
  - **Structuring**
    - **Domains (spheres of action), General Use Case Scenarios, High-level Functions, Classification Criteria for Use Cases, Views for Use Cases**
  - **Documents**
    - **Resources**
  - **Methodology**
    - **Used Concepts / Conventions, Used Verbs**
  - **Process**
    - **Used roles within Use Case creation, Approval Status**

Template for Use Case Compilation

The structure of the use case template is described below, with the additions to IEC/PAS 62559 accented in bold type. The template is structured in two parts, a general part which contains attributes for rough description and a details part which supplements the former with more specific information, especially with regard to IT.

General:

1. Description of the User Requirements of a Function
1.1 Domain Experts
Area of Expertise, Change Description, Approval Status, Version, Date
1.2 Name of Function + **ID and Naming conventions**
Short description (executive summary)
1.3 Scope and Objectives of Function
1.4 Narrative of Function + **optional Diagram in UML**
1.5 Actors: People, Systems, Applications, Databases, Power System other Stakeholder + **Market role and type of interaction**
1.6 Legal Issues: Contracts, Regulations, Policies and Constraints + **Restriction Type, Reference**
Further Constraints: consideration of standards apart from legal documents
1.7 Preconditions and Assumptions, Post conditions and Events + **Events / High-level functions**
Classification + support selection / identification of Use Cases
Notes + comments
2. Drawing or Diagram of Function + **textual description**

Details:

3. Step by Step Analysis of Function
ID and name of activity
Short description of activity
3.1 Preconditions, Assumptions, **Post conditions and events (see general)**
3.2 Steps - Normal Sequence
Configuration, Quality of Service (QoS), Security, Data Management, Constraints and other (catch all) + **Selection of information entries from repository**
3.3 Steps- Alternative, Error Management and / or Maintenance / Backup Sequences
Notes (comments)

Further Conventions/Requirements/Guidelines

It is to be ensured in the compilation of use cases that terms are used consistently and in a uniform manner. This particularly concerns the following entities, but is not limited to them.

- Actors
- Roles
- Classification criteria
- Structuring elements
- Elements of the reference architecture

It is recommended to use established and accepted terms rather than defining new terms. When new, previously non-existent terms are used, it is advisable to submit an application for acceptance of the terms to a use case manager who can decide on the adoption of the new term or can identify an existing term with an equivalent meaning which should be used in its place. Decisions on the adoption of individual terms should not be taken by individuals, but rather by consensus of several responsible persons to ensure an as intersubjective acceptance as possible.

To increase their comprehensibility and acceptance, use cases should be described in such a way that the description is as general and neutral as possible with regard to, for example, technologies, products, companies or projects. This point applies above all to the terms used in the description, for which terms established in the community (if available) should always be preferred to terms which are specific to companies, groups or projects. With regard to international acceptance, it is advisable to follow the guidelines of the IEC.

Furthermore, the stipulation of modeling languages and diagram types is to be recommended, so as to arrive at a uniform description with which, among other factors, easier and more rapid understanding by all those involved can be achieved. It may be possible to disregard the rules and conventions in order to remove obstacles in the compilation of use cases and to accelerate their initial creation, but compliance with those rules and conventions should be ensured by subsequent revision.

Organization and Classification of Use Cases

The organization and classification of use cases are important aspects enhancing the benefits of use case compilation, especially when there is a large number of use cases.

In this context, a distinction is to be made between the structuring of use cases on the one hand, which supports navigation and provides easy access to use cases for potential users, and classification criteria on the other hand, which are predominantly intended to facilitate the identification of use cases.

With regard to structuring, as in a tree diagram, the following hierarchy is proposed:

- Use case scenarios
  - Main function groups
    - Use case
      - Activities

In this regard, a use case is only to be assigned to one domain, while in contrast use case scenarios are to provide a cross-domain ordering characteristic.

The establishment of a structure or hierarchy can however only be usefully begun when several use cases have been compiled. In this respect it is to be ensured that the same degree of detail is maintained on each level wherever possible. An adjustment of the hierarchy will most probably become necessary in the course of time.

Process for General Use Case Compilation and Processing

The process of use case compilation up to elaboration for standardization purposes is divided into five phases which are described in more detail below. An overview of this process is shown in Figure A.1. The energy and IT experts perform the drafting of the use case based on practical experience in several steps. A core team, including in particular a use case manager, provides the necessary conventions and guidelines, and essentially performs an overall consistency and quality assurance function, together with editing of the use case and incorporating it in standardization. Finally, users can search for and identify the classified use cases which are relevant to their concerns.

A more detailed description of the individual roles, their requirements and functions, is presented in the next section, followed by a closer description of the individual phases of the process.

Roles in the Process of Use Case Compilation and Processing

In the following rough description of the roles involved in use case compilation, simplified distinctions are made between energy and IT experts and the core team. Furthermore, only a roughly outlined process is considered in the course of this annex, and the guidelines of the IEC have to be taken into account in further elaboration.

- **Energy expert:** Energy experts possess far-reaching expertise in a particular discipline and contribute substantive information to the use case. They identify new use cases and initiate use case compilation.
- **IT expert:** IT experts possess technical expertise above all in the ICT environment and as a rule only general, high-level knowledge of the energy field. They provide technical details for the use case and can estimate its feasibility in ICT terms.

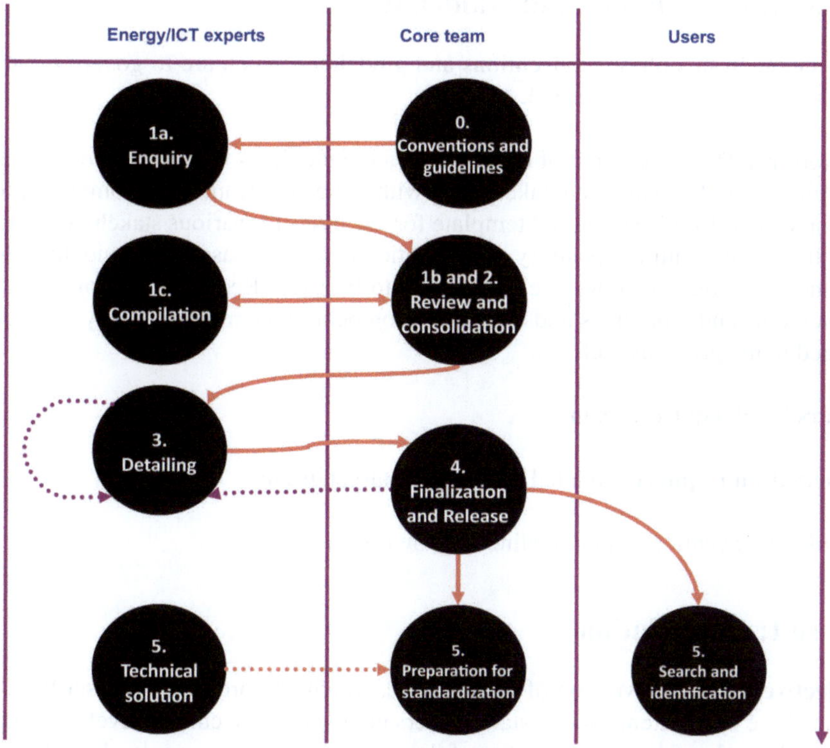

Fig. A.1 Process for use case compilation and processing

- **Core team:** The core team is composed of various experts with broad expertise covering a wide range of disciplines and is entrusted with use case compilation and the conventions involved. With their knowledge of the domain, they are responsible for assuring the quality of the contents and can assess the relevance of use cases. On account of their broad professional expertise spanning several areas of application they can classify the use case in a holistic context (assigning it, for instance, to a professional domain model or a reference architecture). There is at least one specific person in the core team who heads the team as the use case manager and communicates its decisions.

Above and beyond this, the core team includes standardization experts. These experts have outstanding knowledge of established standards and current standardization projects in various standardization organizations, and can assess and classify the relevance of new approaches in this field. Furthermore, they are familiar with the conventions and processes for submission of proposals to standardization organizations. The core team is assisted by use case administrators in the technical maintenance of the use case collection. Support from further persons, e.g. methods experts, is also conceivable.

Phase 0: Conventions and Guidelines

Objective: In this phase, conventions and guidelines which are to govern the use case compilation are established.

Procedure: The elaboration of conventions and guidelines with regard to use case compilation and management takes place within the core team. This comprises, for example, a uniformly structured template for use cases for various stakeholders, the structure of the central repository, classification of the use cases, and guidelines for instance with regard to graphical notations to be used. Recommendations on the conventions and guidelines and the motivation behind them have already been described in the previous sections.

Actors involved: Core team

Information required: Related approaches, target groups

Results: Conventions and guidelines for the use cases

Phase 1: Compilation

Objective: An initial version of the use case, which comprehensibly describes the planned use of a system and the stakeholders involved on a technical level is created in this phase. In addition, the quality of the content of the use case, its thematic relevance to existing use cases and lack of overlaps with them are to be ensured.

Procedure: Compilation or contribution of the initial use case is normally carried out by parties interested in standardization of various aspects of the use case and must be performed in accordance with IEC/PAS 62559 and/or the specific additions based upon it (cf. section entitled Template for use case compilation).

This first version of the use case is established by energy experts who identify the aims of the actors which are to be achieved by the use case. The energy expert describes the use case on the basis of a concrete example (ideally motivated by circumstances in practice), which should be kept as easily comprehensible as possible. The use case outline (step 1a) created up to this point is proposed to the core team which conducts a review and decides on the adoption and therefore further elaboration of the use case (step 1b). In this process, the thematic relevance of the use case in the context of the objective is examined and use cases with the same content are identified (with support by a software tool where appropriate). Should the aim dealt with by the use case be considered irrelevant or should use cases with the same or similar content exist, the use case managers can reject the use case or require sharpening up of the topic.

If the response of the core team is favorable, the use case is further elaborated on a technical basis. This comprises the identification of the actors involved in the use case and their unequivocal definition (if they are not as yet generally known).

Furthermore, restrictions (requirements) and references under which the use case takes place, and conditions and assumptions which must be fulfilled before the use case is performed, are defined.

Actors involved: Energy experts, core team

Information required: Concrete use case with need for technical support, list of defined actors, central repository

Results: Technical part of the use case (aims of the system, description of the use case, actors involved, restrictions and requirements, conditions and assumptions)

Phase 2: Review and Consolidation

Objective: In the consolidation phase, the terminology used and compliance with conventions etc. are reviewed, and amendments or additions made as necessary. Finally, a classification of the use case and its location in the overall context take place, so that efficient evaluation and identification are possible.

Procedure: The terminology, conventions etc. used in the technical part of the use case are reviewed and if necessary adjusted for conformity with the specified terminology. Terms not yet available in the central repository which represent essential new concepts in the use case can be included in the central repository if required.

The use case is then enhanced as required with diagrams as stipulated (e.g. particular UML diagrams - see also [17]) and existing, non-conforming diagrams are converted into a form which complies with the guidelines. This is intended to ensure the consistency of graphical notations throughout the collection of use cases and to contribute to ease of understanding for the readers.

Furthermore, possible references to laws or regulations and to standards and corresponding working groups which are relevant in the context of the use case are added at this stage.

The core team in its role of dealing with cross-cutting issues then finally classifies the use case on the basis of specified classification criteria (cf. section entitled Organization and classification of use cases).

Actors involved: Core team
Information required: Initial version of the use case from energy experts, central repository, list of actors
Results: Revised version of the technical part of the use case (adjusted terminology, references, classification)

Phase 3: Detailing

Objective: Addition of further details and workflows to the use case.

Procedure: The use case developed up to this point is provided with further detail in this phase. This is done by IT experts, who further refine the use case by adding and describing activities. These activities represent the individual steps of the use case, which have to be performed in a defined sequence to achieve the intended aim.

Over and above this, alternative sequences of the activities which may, for example, occur in the event of a fault, are also identified in this phase. Conditions and assumptions which are essential for the performance of the activity are also defined for the individual steps. The identification of the individual steps and alternative sequences may result in the necessity of introducing further steps, and so this detailing phase may have to be repeated, leading to an incremental refinement of the use case.

Actors involved: IT experts

Information required: Accepted, revised version of the use case from the use case managers, central repository, list of actors

Results: Detailed version of the use case (description of the individual activities in the use case)

Phase 4: Finalization and Release

Objective: Review of the use case and decision on further processing or release to go forward into standardization or for preparation for standardization and for the interested professional public.

Procedure: The use case, now detailed, is reviewed by the core team in this phase with regard to completeness, and once again for compliance with conventions. The terminology, conventions etc. used are checked and if necessary adjusted for conformity with the specified terminology. Terms not yet available in the central repository which represent essential new concepts in the use case can be included in the central repository if required.

The use case is enhanced as required with diagrams as stipulated (e.g. particular UML diagrams) and existing, non-conforming diagrams are converted into a form which complies with the guidelines. This is intended to ensure the consistency of graphical notations throughout the collection of use cases and to contribute to ease of understanding for the readers.

Furthermore, possible references to laws or regulations and to standards and corresponding working groups which are relevant in the context of the activities in the use case are added at this stage.

Finally, the use case is released by the use case managers for the further standardization process or the preparation process for standardization, or the use case is sent back for further processing (phase 3), so as to arrange for any changes by the IT experts.

Actors involved: Core team (especially use case manager)

Information required: Detailed version of the use case from the IT experts, central repository, list of actors

Results: Final version of the use case for the further standardization processes

Phase 5: Preparation for Standardization/Use/Identification

Objective: Extension of a use case for the relevant standardization organization and subsequent use for implementation. Additionally, identification of described use cases.

Procedure: From the finalized use cases and from any solutions which have already been developed by the energy and IT experts, the core team identifies parts to be standardized from which corresponding conceptual designs (e.g. for data modeling, interfaces, etc.) are derived, or parts to be standardized are proposed by energy and IT experts.

The standardization expert is given the task to prepare these IT parts and relevant parts of the use case and technical solution concept resulting from the requirements documented for a specific standardization organization. The use case is then prepared in accordance with the requirements of the standardization organization and goes forward into the further standardization process.

Users of the use cases can locate use cases relevant to them on the basis of the classification and use or implement the use cases and/or the resulting standards.

In the context of further use of standards, ranging up to implementation, the subsequent compilation of dedicated test use cases and standard profiles is conceivable. These are to be appropriately detailed to permit technical implementation or testing for interoperability where necessary.

Actors involved: Core team (especially standardization expert), energy and IT experts, use case users Information required: Final version of the use case from the core team, standardization organization to which the use case is to be submitted, and organization?s requirements for use cases

Results: Version of the use case prepared for the relevant standardization organization.

Appendix B
Basic Message Structure in XML Schema

Listing B.1 Basic message structure in XML Schema

```
<?xml version="1.0" encoding="utf-8"?>
<!-- Common Message Specification for ESB Integration via IEC 61968 -->
<xs:schema xmlns="http://www.iec.ch/TC57/2008/schema/message"
xmlns:xs="http://www.w3.org/2001/XMLSchema"
targetNamespace="http://www.iec.ch/TC57/2008/schema/message"
elementFormDefault="qualified" attributeFormDefault="unqualified"
version="1.0.0">
  <xs:complexType name="RequestType">
    <xs:annotation>
      <xs:documentation>Request type definition</xs:documentation>
    </xs:annotation>
    <xs:sequence>
      <xs:annotation>
        <xs:documentation>Request package is typically used to
        supply parameters for 'get' requests</xs:documentation>
      </xs:annotation>
      <xs:element name="StartTime" type="xs:dateTime"
      minOccurs="0">
        <xs:annotation>
          <xs:documentation>Start time of
          interest</xs:documentation>
        </xs:annotation>
      </xs:element>
      <xs:element name="EndTime" type="xs:dateTime" minOccurs="0">
        <xs:annotation>
          <xs:documentation>End time of interest</xs:documentation>
        </xs:annotation>
      </xs:element>
      <xs:element name="Option" type="xs:string" minOccurs="0"
      maxOccurs="unbounded">
        <xs:annotation>
          <xs:documentation>Request type
          specialization</xs:documentation>
        </xs:annotation>
      </xs:element>
      <xs:element name="ID" type="xs:string" minOccurs="0"
      maxOccurs="unbounded">
        <xs:annotation>
          <xs:documentation>Object ID for
          request</xs:documentation>
        </xs:annotation>
      </xs:element>
      <xs:any namespace="##other" processContents="lax"
```

```
      minOccurs="0" maxOccurs="unbounded">
        <xs:annotation>
          <xs:documentation>This can be a CIM profile defined as an
          XSD with a CIM-specific namespace</xs:documentation>
        </xs:annotation>
      </xs:any>
    </xs:sequence>
  </xs:complexType>
  <xs:complexType name="ReplyType">
    <xs:annotation>
      <xs:documentation>Reply type definition</xs:documentation>
    </xs:annotation>
    <xs:sequence>
      <xs:annotation>
        <xs:documentation>Reply package is used to confirm success
        or report errors</xs:documentation>
      </xs:annotation>
      <xs:element name="ReplyCode" type="xs:string">
        <xs:annotation>
          <xs:documentation>Reply code: OK or application defined
          error code, such an IEC 61968-9
          enumeration</xs:documentation>
        </xs:annotation>
      </xs:element>
      <xs:element name="Error" type="ErrorType" minOccurs="0"
      maxOccurs="unbounded">
        <xs:annotation>
          <xs:documentation>Reply details describing one or more
          errors</xs:documentation>
        </xs:annotation>
      </xs:element>
      <xs:element name="ID" type="xs:string" minOccurs="0"
      maxOccurs="unbounded">
        <xs:annotation>
          <xs:documentation>Resulting transaction ID (usually
          consequence of create)</xs:documentation>
        </xs:annotation>
      </xs:element>
      <xs:any namespace="##other" processContents="lax"
      minOccurs="0" maxOccurs="unbounded" />
    </xs:sequence>
  </xs:complexType>
  <xs:complexType name="PayloadType">
    <xs:annotation>
      <xs:documentation>Payload container</xs:documentation>
    </xs:annotation>
    <xs:sequence>
      <xs:choice>
        <xs:any namespace="##other" processContents="skip"
        minOccurs="0" maxOccurs="unbounded">
          <xs:annotation>
            <xs:documentation>For XML payloads, usually CIM
            profiles defined using an XSD in a profile-specific
            namespace</xs:documentation>
          </xs:annotation>
        </xs:any>
        <xs:element name="Compressed" type="xs:string"
        minOccurs="0">
          <xs:annotation>
            <xs:documentation>For compressed and/or binary,
            uuencoded payloads</xs:documentation>
          </xs:annotation>
        </xs:element>
      </xs:choice>
      <xs:element name="format" type="xs:string" minOccurs="0">
        <xs:annotation>
          <xs:documentation>Hint as to format of payload, e.g. XML,
```

```
        RDF, SVF, BINARY, PDF, ...</xs:documentation>
      </xs:annotation>
    </xs:element>
  </xs:sequence>
</xs:complexType>
<xs:complexType name="ReplayDetectionType">
  <xs:annotation>
    <xs:documentation>Used to detect and prevent replay
    attacks</xs:documentation>
  </xs:annotation>
  <xs:sequence>
    <xs:element name="Nonce" type="xs:string" />
    <xs:element name="Created" type="xs:dateTime" />
  </xs:sequence>
</xs:complexType>
<xs:complexType name="UserType">
  <xs:annotation>
    <xs:documentation>User type definition</xs:documentation>
  </xs:annotation>
  <xs:sequence>
    <xs:element name="UserID" type="xs:string">
      <xs:annotation>
        <xs:documentation>User identifier</xs:documentation>
      </xs:annotation>
    </xs:element>
    <xs:element name="Organization" type="xs:string">
      <xs:annotation>
        <xs:documentation>User parent organization
        identifier</xs:documentation>
      </xs:annotation>
    </xs:element>
  </xs:sequence>
</xs:complexType>
<xs:complexType name="HeaderType">
  <xs:annotation>
    <xs:documentation>Message header type
    definition</xs:documentation>
  </xs:annotation>
  <xs:sequence>
    <xs:annotation>
      <xs:documentation>Message header contains control and
      descriptive information about the
      message.</xs:documentation>
    </xs:annotation>
    <xs:element name="Verb">
      <xs:annotation>
        <xs:documentation>This enumerated list of verbs that can
        be used to form message types in compliance with the IEC
        61968 standard.</xs:documentation>
      </xs:annotation>
      <xs:simpleType>
        <xs:restriction base="xs:string">
          <xs:enumeration value="cancel" />
          <xs:enumeration value="canceled" />
          <xs:enumeration value="change" />
          <xs:enumeration value="changed" />
          <xs:enumeration value="create" />
          <xs:enumeration value="created" />
          <xs:enumeration value="close" />
          <xs:enumeration value="closed" />
          <xs:enumeration value="delete" />
          <xs:enumeration value="deleted" />
          <xs:enumeration value="get" />
          <xs:enumeration value="show" />
          <xs:enumeration value="reply" />
          <xs:enumeration value="subscribe" />
          <xs:enumeration value="unsubscribe" />
```

```
        <xs:enumeration value="execute" />
        <xs:enumeration value="report" />
        <xs:enumeration value="stop" />
        <xs:enumeration value="terminate" />
      </xs:restriction>
    </xs:simpleType>
  </xs:element>
  <xs:element name="Noun" type="xs:string">
    <xs:annotation>
      <xs:documentation>The Noun of the Control Area identifies
      the main subject of the message type, typically a real
      world object defined in the CIM.</xs:documentation>
    </xs:annotation>
  </xs:element>
  <xs:element name="Revision" type="xs:string" minOccurs="0">
    <xs:annotation>
      <xs:documentation>Revision level of the message
      type.</xs:documentation>
    </xs:annotation>
  </xs:element>
  <xs:element name="ReplayDetection" type="ReplayDetectionType"
  minOccurs="0" />
  <xs:element name="Context" minOccurs="0">
    <xs:annotation>
      <xs:documentation>Intended context for information
      usage</xs:documentation>
    </xs:annotation>
    <xs:simpleType>
      <xs:restriction base="xs:string">
        <xs:enumeration value="PRODUCTION" />
        <xs:enumeration value="TESTING" />
        <xs:enumeration value="DEVELOPMENT" />
        <xs:enumeration value="STUDY" />
        <xs:enumeration value="TRAINING" />
      </xs:restriction>
    </xs:simpleType>
  </xs:element>
  <xs:element name="Timestamp" type="xs:dateTime"
  minOccurs="0">
    <xs:annotation>
      <xs:documentation>Application level relevant time and
      date for when this instance of the message type was
      produced. This is not intended to be used by middleware
      for message management.</xs:documentation>
    </xs:annotation>
  </xs:element>
  <xs:element name="Source" type="xs:string" minOccurs="0">
    <xs:annotation>
      <xs:documentation>Source person or system that publishes
      the message</xs:documentation>
    </xs:annotation>
  </xs:element>
  <xs:element name="AsyncReplyFlag" type="xs:boolean"
  minOccurs="0">
    <xs:annotation>
      <xs:documentation>Indicates whether or not reply should
      be asynchronous</xs:documentation>
    </xs:annotation>
  </xs:element>
  <xs:element name="ReplyAddress" type="xs:string"
  minOccurs="0">
    <xs:annotation>
      <xs:documentation>Address to be used for asynchronous
      replies</xs:documentation>
    </xs:annotation>
  </xs:element>
  <xs:element name="AckRequired" type="xs:boolean"
```

```
      minOccurs="0">
        <xs:annotation>
          <xs:documentation>Indicates whether or not an
          acknowledgement is required</xs:documentation>
        </xs:annotation>
      </xs:element>
      <xs:element name="User" type="UserType" minOccurs="0">
        <xs:annotation>
          <xs:documentation>User information of the
          sender</xs:documentation>
        </xs:annotation>
      </xs:element>
      <xs:element name="MessageID" type="xs:string" minOccurs="0">
        <xs:annotation>
          <xs:documentation>Unique message ID to be used for
          tracking messages</xs:documentation>
        </xs:annotation>
      </xs:element>
      <xs:element name="CorrelationID" type="xs:string"
      minOccurs="0">
        <xs:annotation>
          <xs:documentation>ID to be used by applications for
          correlating replies</xs:documentation>
        </xs:annotation>
      </xs:element>
      <xs:element name="Comment" type="xs:string" minOccurs="0">
        <xs:annotation>
          <xs:documentation>Optional comment</xs:documentation>
        </xs:annotation>
      </xs:element>
      <xs:element name="Property" type="MessageProperty"
      minOccurs="0" maxOccurs="unbounded">
        <xs:annotation>
          <xs:documentation>Message properties can be used to
          identify information needed for extended routing and
          filtering capabilities</xs:documentation>
        </xs:annotation>
      </xs:element>
      <xs:any namespace="##other" processContents="lax"
      minOccurs="0" maxOccurs="unbounded" />
    </xs:sequence>
  </xs:complexType>
  <xs:element name="Message" type="MessageType">
    <xs:annotation>
      <xs:documentation>Common IEC 61968 Message
      Definition</xs:documentation>
    </xs:annotation>
  </xs:element>
  <xs:complexType name="MessageProperty">
    <xs:annotation>
      <xs:documentation>Message properties can be used for extended
      routing and filtering</xs:documentation>
    </xs:annotation>
    <xs:sequence>
      <xs:element name="Name" type="xs:string" />
      <xs:element name="Value" type="xs:string" minOccurs="0" />
    </xs:sequence>
  </xs:complexType>
  <xs:element name="RequestMessage" type="RequestMessageType">
    <xs:annotation>
      <xs:documentation>Request message
      structure</xs:documentation>
    </xs:annotation>
  </xs:element>
  <xs:element name="ResponseMessage" type="ResponseMessageType">
    <xs:annotation>
      <xs:documentation>Response message
```

```
        structure</xs:documentation>
      </xs:annotation>
    </xs:element>
    <xs:element name="FaultMessage" type="FaultMessageType">
      <xs:annotation>
        <xs:documentation>Fault message structure</xs:documentation>
      </xs:annotation>
    </xs:element>
    <xs:element name="EventMessage" type="EventMessageType">
      <xs:annotation>
        <xs:documentation>Event message structure</xs:documentation>
      </xs:annotation>
    </xs:element>
    <xs:complexType name="MessageType">
      <xs:annotation>
        <xs:documentation>Generic Message Type</xs:documentation>
      </xs:annotation>
      <xs:sequence>
        <xs:element name="Header" type="HeaderType" />
        <xs:element name="Request" type="RequestType"
        minOccurs="0" />
        <xs:element name="Reply" type="ReplyType" minOccurs="0" />
        <xs:element name="Payload" type="PayloadType"
        minOccurs="0" />
      </xs:sequence>
    </xs:complexType>
    <xs:complexType name="RequestMessageType">
      <xs:annotation>
        <xs:documentation>Request Message Type, which will typically
        result in a ResponseMessage to be returned. This isn
        typically used to initiate a transaction or a query
        request.</xs:documentation>
      </xs:annotation>
      <xs:sequence>
        <xs:element name="Header" type="HeaderType" />
        <xs:element name="Request" type="RequestType"
        minOccurs="0" />
        <xs:element name="Payload" type="PayloadType"
        minOccurs="0" />
      </xs:sequence>
    </xs:complexType>
    <xs:complexType name="ResponseMessageType">
      <xs:annotation>
        <xs:documentation>Response MessageType, typically used to
        reply to a RequestMessage</xs:documentation>
      </xs:annotation>
      <xs:sequence>
        <xs:element name="Header" type="HeaderType" />
        <xs:element name="Reply" type="ReplyType" />
        <xs:element name="Payload" type="PayloadType"
        minOccurs="0" />
      </xs:sequence>
    </xs:complexType>
    <xs:complexType name="FaultMessageType">
      <xs:annotation>
        <xs:documentation>Fault Messsage Type, which is used in cases
        where the incoming message (including the header) can not be
        parsed</xs:documentation>
      </xs:annotation>
      <xs:sequence>
        <xs:element name="Reply" type="ReplyType" />
      </xs:sequence>
    </xs:complexType>
    <xs:complexType name="EventMessageType">
      <xs:annotation>
        <xs:documentation>Event Message Type, which is used to
        indicate a condition of potential
```

```
      interest.</xs:documentation>
    </xs:annotation>
    <xs:sequence>
      <xs:element name="Header" type="HeaderType" />
      <xs:element name="Request" type="RequestType"
      minOccurs="0" />
      <xs:element name="Payload" type="PayloadType"
      minOccurs="0" />
    </xs:sequence>
  </xs:complexType>
  <xs:complexType name="ErrorType">
    <xs:annotation>
      <xs:documentation>Error Structure</xs:documentation>
    </xs:annotation>
    <xs:sequence>
      <xs:element name="level">
        <xs:annotation>
          <xs:documentation>Severity level</xs:documentation>
        </xs:annotation>
        <xs:simpleType>
          <xs:restriction base="xs:string">
            <xs:enumeration value="INFORM" />
            <xs:enumeration value="WARNING" />
            <xs:enumeration value="FATAL" />
            <xs:enumeration value="CATASTROPHIC" />
          </xs:restriction>
        </xs:simpleType>
      </xs:element>
      <xs:element name="code" type="xs:string">
        <xs:annotation>
          <xs:documentation>Application defined error
          code</xs:documentation>
        </xs:annotation>
      </xs:element>
      <xs:element name="xpath" type="xs:QName" minOccurs="0">
        <xs:annotation>
          <xs:documentation>XPath expression to identify specific
          XML element</xs:documentation>
        </xs:annotation>
      </xs:element>
      <xs:element name="details" type="xs:string" minOccurs="0">
        <xs:annotation>
          <xs:documentation>Free form text description of
          error</xs:documentation>
        </xs:annotation>
      </xs:element>
    </xs:sequence>
  </xs:complexType>
</xs:schema>
```

Appendix C
Customer Example Schema

Listing C.1 CustomerExample XML Schema generated from CIMTool

```
<?xml version="1.0" encoding="utf-8"?>
<xs:schema xmlns:xs="http://www.w3.org/2001/XMLSchema"
xmlns:a="http://langdale.com.au/2005/Message#"
xmlns:sawsdl="http://www.w3.org/ns/sawsdl"
xmlns="http://langdale.com.au/2005/Message#"
xmlns:m="http://iec.ch/TC57/2007/CustomerExample#"
targetNamespace="http://iec.ch/TC57/2007/CustomerExample#"
elementFormDefault="qualified" attributeFormDefault="unqualified">
  <xs:element name="CustomerExamples" type="m:CustomerExamples" />
  <xs:complexType name="CustomerExamples">
    <xs:sequence>
      <xs:element name="Customer" type="m:Customer" minOccurs="0"
      maxOccurs="unbounded" />
    </xs:sequence>
  </xs:complexType>
  <xs:complexType name="Customer"
  sawsdl:modelReference="http://iec.ch/TC57/CIM-generic#Customer">
    <xs:annotation>
      <xs:documentation>Organisation receiving services from
      ServiceSupplier.</xs:documentation>
    </xs:annotation>
    <xs:sequence>
      <xs:element name="mRID" type="xs:string"
      sawsdl:modelReference="http://iec.ch/TC57/CIM-generic#
                                  IdentifiedObject.mRID">

        <xs:annotation>
          <xs:documentation>A Model Authority issues mRIDs. Given
          that each Model Authority has a unique id and this id is
          part of the mRID, then the mRID is globally
          unique.</xs:documentation>
        </xs:annotation>
      </xs:element>
      <xs:element name="name" type="xs:string"
      sawsdl:modelReference="http://iec.ch/TC57/CIM-generic#
                                  IdentifiedObject.name">

        <xs:annotation>
          <xs:documentation>The name is a free text human readable
          name of the object. It may be non unique and may not
          correlate to a naming hierarchy.</xs:documentation>
        </xs:annotation>
      </xs:element>
      <xs:element name="postalAddress"
```

```
sawsdl:modelReference="http://iec.ch/TC57/CIM-generic#
                          Organisation.postalAddress">

  <xs:annotation>
    <xs:documentation>Postal address, potentially different
    than 'streetAddress' (e.g., another
    city).</xs:documentation>
  </xs:annotation>
  <xs:complexType sawsdl:modelReference=
                "http://iec.ch/TC57/CIM-generic#
        PostalAddress">

    <xs:sequence>
      <xs:element name="postalCode" type="xs:string"
      sawsdl:modelReference="http://iec.ch/TC57/CIM-generic#
                              PostalAddress.postalCode">

        <xs:annotation>
          <xs:documentation>Postal code for the
          address.</xs:documentation>
        </xs:annotation>
      </xs:element>
      <xs:element name="streetDetail"
      sawsdl:modelReference="http://iec.ch/TC57/CIM-generic#
                              PostalAddress.streetDetail">

        <xs:annotation>
          <xs:documentation>Street detail.</xs:documentation>
        </xs:annotation>
        <xs:complexType sawsdl:modelReference=
                  "http://iec.ch/TC57/CIM-generic#StreetDetail">

          <xs:sequence>
            <xs:element name="name" type="xs:string"
            sawsdl:modelReference="http://iec.ch/TC57/CIM-generic#
                                    StreetDetail.name">

              <xs:annotation>
                <xs:documentation>Name of the
                street.</xs:documentation>
              </xs:annotation>
            </xs:element>
            <xs:element name="number" type="xs:string"
            sawsdl:modelReference="http://iec.ch/TC57/CIM-generic#
                                    StreetDetail.number">

              <xs:annotation>
                <xs:documentation>Designator of the specific
                location on the street.</xs:documentation>
              </xs:annotation>
            </xs:element>
          </xs:sequence>
        </xs:complexType>
      </xs:element>
      <xs:element name="townDetail"
      sawsdl:modelReference="http://iec.ch/TC57/CIM-generic#
                                    PostalAddress.townDetail">

        <xs:annotation>
          <xs:documentation>Town detail.</xs:documentation>
        </xs:annotation>
        <xs:complexType sawsdl:modelReference=
                          "http://iec.ch/TC57/CIM-generic#TownDetail">

          <xs:sequence>
            <xs:element name="name" type="xs:string"
            sawsdl:modelReference="http://iec.ch/TC57/CIM-generic#
```

```
                                                TownDetail.name">

                    <xs:annotation>
                      <xs:documentation>Town
                      name.</xs:documentation>
                    </xs:annotation>
                  </xs:element>
                </xs:sequence>
              </xs:complexType>
            </xs:element>
          </xs:sequence>
        </xs:complexType>
      </xs:element>
    </xs:sequence>
  </xs:complexType>
</xs:schema>
```

Appendix D
EndDeviceEvent Message Structure

Listing D.1 EndDeviceEvent message Structure

```
<?xml version="1.0" encoding="utf-8"?>
<xs:schema xmlns:xs="http://www.w3.org/2001/XMLSchema"
xmlns:a="http://langdale.com.au/2005/Message#"
xmlns:sawsdl="http://www.w3.org/ns/sawsdl"
targetNamespace="http://iec.ch/TC57/2009/EndDeviceEvents#"
elementFormDefault="qualified" attributeFormDefault="unqualified"
xmlns="http://langdale.com.au/2005/Message#"
xmlns:m="http://iec.ch/TC57/2009/EndDeviceEvents#">
  <xs:element name="EndDeviceEvents" type="m:EndDeviceEvents" />
  <xs:complexType name="EndDeviceEvents">
    <xs:sequence>
      <xs:element name="EndDeviceEvent" type="m:EndDeviceEvent"
      minOccurs="0" maxOccurs="unbounded" />
    </xs:sequence>
  </xs:complexType>
  <xs:complexType name="EndDeviceEvent"
  sawsdl:modelReference="http://iec.ch/TC57/CIM generic#
                        EndDeviceEvent">

    <xs:sequence>
      <xs:element name="mRID" minOccurs="0" maxOccurs="1"
      type="xs:string"
      sawsdl:modelReference="http://iec.ch/TC57/CIM-generic#
                            IdentifiedObject.mRID" />
      <xs:element name="category" minOccurs="1" maxOccurs="1"
      type="xs:string"
      sawsdl:modelReference="http://iec.ch/TC57/CIM-generic#
                            ActivityRecord.category" />
      <xs:element name="createdDateTime" minOccurs="1"
      maxOccurs="1" type="xs:dateTime"
      sawsdl:modelReference="http://iec.ch/TC57/CIM-generic#
                            ActivityRecord.createdDateTime" />
      <xs:element name="description" minOccurs="0" maxOccurs="1"
      type="xs:string"
      sawsdl:modelReference="http://iec.ch/TC57/CIM-generic#
                            IdentifiedObject.description" />
      <xs:element name="reason" minOccurs="0" maxOccurs="1"
      type="xs:string"
      sawsdl:modelReference="http://iec.ch/TC57/CIM-generic#
                            ActivityRecord.reason" />
      <xs:element name="severity" minOccurs="0" maxOccurs="1"
      type="xs:string"
      sawsdl:modelReference="http://iec.ch/TC57/CIM-generic#
                            ActivityRecord.severity" />
```

```
      <xs:element name="userID" minOccurs="0" maxOccurs="1"
      type="xs:string"
      sawsdl:modelReference="http://iec.ch/TC57/CIM-generic#
                             EndDeviceEvent.userID" />
      <xs:element name="Assets" minOccurs="1" maxOccurs="1"
      sawsdl:modelReference="http://iec.ch/TC57/CIM-generic#
                             ActivityRecord.Assets">

        <xs:complexType sawsdl:modelReference=
                        "http://iec.ch/TC57/CIM-generic#Asset">

          <xs:sequence>
            <xs:element name="mRID" minOccurs="1" maxOccurs="1"
            type="xs:string"
            sawsdl:modelReference="http://iec.ch/TC57/CIM-generic#
                                   IdentifiedObject.mRID" />
          </xs:sequence>
        </xs:complexType>
      </xs:element>
      <xs:element name="status" minOccurs="0" maxOccurs="1"
      sawsdl:modelReference="http://iec.ch/TC57/CIM-generic#
                             ActivityRecord.status">

        <xs:complexType sawsdl:modelReference=
                        "http://iec.ch/TC57/CIM-generic#Status">

          <xs:sequence>
            <xs:element name="dateTime" minOccurs="0" maxOccurs="1"
            type="xs:dateTime"
            sawsdl:modelReference="http://iec.ch/TC57/CIM-generic#
                                       Status.dateTime" />
            <xs:element name="reason" minOccurs="0" maxOccurs="1"
            type="xs:string"
            sawsdl:modelReference="http://iec.ch/TC57/CIM-generic#
                                       Status.reason" />
            <xs:element name="value" minOccurs="1" maxOccurs="1"
            type="xs:string"
            sawsdl:modelReference="http://iec.ch/TC57/CIM-generic#
                                       Status.value" />
          </xs:sequence>
        </xs:complexType>
      </xs:element>
    </xs:sequence>
  </xs:complexType>
</xs:schema>
```

Appendix E
Topology Example

Listing E.1 Complete RDF serialization for the topology example

```xml
<?xml version="1.0" encoding="iso-8859-1"?>
<rdf:RDF xmlns:cim="http://iec.ch/TC57/2008/CIM-schema-cim13#"
xmlns:rdf="http://www.w3.org/1999/02/22-rdf-syntax-ns#">
  <cim:Substation rdf:ID="A_Z">
    <cim:IdentifiedObject.aliasName>Substation
    Z</cim:IdentifiedObject.aliasName>
    <cim:IdentifiedObject.name>Substation
    110kV/20kV</cim:IdentifiedObject.name>
  </cim:Substation>
  <cim:PowerTransformer rdf:ID="A_EF">
    <cim:Equipment.MemberOf_EquipmentContainer rdf:resource="#A_Z" />
    <cim:IdentifiedObject.name>Transformer
    EF</cim:IdentifiedObject.name>
    <cim:IdentifiedObject.mRID>2</cim:IdentifiedObject.mRID>
  </cim:PowerTransformer>
  <cim:TransformerWinding rdf:ID="A_EF1">
    <cim:TransformerWinding.connectionType rdf:resource=
    "http://iec.ch/TC57/2008/CIM-schema-cim13#WindingConnection.Y" />
    <cim:TransformerWinding.ratedS>
    40</cim:TransformerWinding.ratedS>
    <cim:TransformerWinding.ratedU>
    110</cim:TransformerWinding.ratedU>
    <cim:TransformerWinding.windingType rdf:resource=
    "http://iec.ch/TC57/2008/CIM-schema-cim13#WindingType.primary" />
    <cim:TransformerWinding.MemberOf_PowerTransformer rdf:resource="#A_EF" />
    <cim:IdentifiedObject.mRID>3</cim:IdentifiedObject.mRID>
  </cim:TransformerWinding>
  <cim:TransformerWinding rdf:ID="A_EF2">
    <cim:TransformerWinding.connectionType rdf:resource=
    "http://iec.ch/TC57/2008/CIM-schema-cim13#WindingConnection.Y" />
    <cim:TransformerWinding.windingType rdf:resource=
    "http://iec.ch/TC57/2008/CIM-schema-cim13#WindingType.secondary" />
    <cim:TransformerWinding.MemberOf_PowerTransformer rdf:resource="#A_EF" />
    <cim:TransformerWinding.ratedU>
    20</cim:TransformerWinding.ratedU>
  </cim:TransformerWinding>
  <cim:Terminal rdf:ID="A_T1">
    <cim:Terminal.ConductingEquipment rdf:resource="#A_EF2" />
    <cim:Terminal.ConnectivityNode rdf:resource="#A_CN1" />
  </cim:Terminal>
  <cim:ConnectivityNode rdf:ID="A_CN1"></cim:ConnectivityNode>
  <cim:Terminal rdf:ID="A_T2">
    <cim:Terminal.ConductingEquipment rdf:resource="#A_H" />
    <cim:Terminal.ConnectivityNode rdf:resource="#A_CN1" />
```

```
</cim:Terminal>
<cim:BusbarSection rdf:ID="A_H">
  <cim:Equipment.MemberOf_EquipmentContainer rdf:resource="#A_Z" />
  <cim:IdentifiedObject.name>Busbar H</cim:IdentifiedObject.name>
  <cim:IdentifiedObject.mRID>8</cim:IdentifiedObject.mRID>
</cim:BusbarSection>
<cim:Terminal rdf:ID="A_T3">
  <cim:Terminal.ConductingEquipment rdf:resource="#A_A1" />
  <cim:Terminal.ConnectivityNode rdf:resource="#A_CN1" />
</cim:Terminal>
<cim:Line rdf:ID="A_A">
  <cim:IdentifiedObject.name>Line A</cim:IdentifiedObject.name>
  <cim:IdentifiedObject.mRID>34</cim:IdentifiedObject.mRID>
</cim:Line>
<cim:ACLineSegment rdf:ID="A_A1">
  <cim:Conductor.length>2500</cim:Conductor.length>
  <cim:Conductor.r>0.3125</cim:Conductor.r>
  <cim:Conductor.x>0.28</cim:Conductor.x>
  <cim:Conductor.bch>235.45</cim:Conductor.bch>
  <cim:Equipment.MemberOf_EquipmentContainer rdf:resource="#A_A" />
  <cim:IdentifiedObject.name>Line A1</cim:IdentifiedObject.name>
  <cim:IdentifiedObject.mRID>10</cim:IdentifiedObject.mRID>
</cim:ACLineSegment>
<cim:Terminal rdf:ID="A_T5">
  <cim:Terminal.ConductingEquipment rdf:resource="#A_A1" />
  <cim:Terminal.ConnectivityNode rdf:resource="#A_CN2" />
</cim:Terminal>
<cim:ConnectivityNode rdf:ID="A_CN2"></cim:ConnectivityNode>
<cim:Terminal rdf:ID="A_T6">
  <cim:Terminal.ConductingEquipment rdf:resource="#A_A2" />
  <cim:Terminal.ConnectivityNode rdf:resource="#A_CN2" />
</cim:Terminal>
<cim:ACLineSegment rdf:ID="A_A2">
  <cim:Conductor.length>2500</cim:Conductor.length>
  <cim:Conductor.r>0.3125</cim:Conductor.r>
  <cim:Conductor.x>0.28</cim:Conductor.x>
  <cim:Conductor.bch>235.45</cim:Conductor.bch>
  <cim:Equipment.MemberOf_EquipmentContainer rdf:resource="#A_A" />
  <cim:IdentifiedObject.name>Line A2</cim:IdentifiedObject.name>
  <cim:IdentifiedObject.mRID>14</cim:IdentifiedObject.mRID>
</cim:ACLineSegment>
<cim:Terminal rdf:ID="A_T8">
  <cim:Terminal.ConductingEquipment rdf:resource="#A_A2" />
  <cim:Terminal.ConnectivityNode rdf:resource="#A_CN3" />
</cim:Terminal>
<cim:ConnectivityNode rdf:ID="A_CN3"></cim:ConnectivityNode>
<cim:Terminal rdf:ID="A_T7">
  <cim:Terminal.ConductingEquipment rdf:resource="#A_K" />
  <cim:Terminal.ConnectivityNode rdf:resource="#A_CN3" />
</cim:Terminal>
<cim:BusbarSection rdf:ID="A_K">
  <cim:IdentifiedObject.name>Busbar K</cim:IdentifiedObject.name>
  <cim:IdentifiedObject.mRID>18</cim:IdentifiedObject.mRID>
</cim:BusbarSection>
<cim:Terminal rdf:ID="A_T9">
  <cim:Terminal.ConductingEquipment rdf:resource="#A_B" />
  <cim:Terminal.ConnectivityNode rdf:resource="#A_CN3" />
</cim:Terminal>
<cim:ACLineSegment rdf:ID="A_B">
  <cim:Conductor.length>3000</cim:Conductor.length>
  <cim:Conductor.r>0.375</cim:Conductor.r>
  <cim:Conductor.x>0.336</cim:Conductor.x>
  <cim:Conductor.bch>282.75</cim:Conductor.bch>
  <cim:IdentifiedObject.name>Line B</cim:IdentifiedObject.name>
  <cim:IdentifiedObject.mRID>20</cim:IdentifiedObject.mRID>
</cim:ACLineSegment>
<cim:Terminal rdf:ID="A_T10">
```

```
    <cim:Terminal.ConductingEquipment rdf:resource="#A_B" />
    <cim:Terminal.ConnectivityNode rdf:resource="#A_CN4" />
  </cim:Terminal>
  <cim:ConnectivityNode rdf:ID="A_CN4"></cim:ConnectivityNode>
  <cim:Terminal rdf:ID="A_T11">
    <cim:Terminal.ConductingEquipment rdf:resource="#A_L" />
    <cim:Terminal.ConnectivityNode rdf:resource="#A_CN4" />
  </cim:Terminal>
  <cim:BusbarSection rdf:ID="A_L">
    <cim:IdentifiedObject.name>Busbar L</cim:IdentifiedObject.name>
    <cim:IdentifiedObject.mRID>24</cim:IdentifiedObject.mRID>
  </cim:BusbarSection>
  <cim:Terminal rdf:ID="A_T12">
    <cim:Terminal.ConductingEquipment rdf:resource="#A_C" />
    <cim:Terminal.ConnectivityNode rdf:resource="#A_CN4" />
  </cim:Terminal>
  <cim:ACLineSegment rdf:ID="A_C">
    <cim:Conductor.length>300</cim:Conductor.length>
    <cim:Conductor.r>0.0375</cim:Conductor.r>
    <cim:Conductor.x>0.0336</cim:Conductor.x>
    <cim:Conductor.bch>28.275</cim:Conductor.bch>
    <cim:IdentifiedObject.name>Line C</cim:IdentifiedObject.name>
    <cim:IdentifiedObject.mRID>26</cim:IdentifiedObject.mRID>
  </cim:ACLineSegment>
  <cim:Terminal rdf:ID="A_T13">
    <cim:Terminal.ConductingEquipment rdf:resource="#A_C" />
    <cim:Terminal.ConnectivityNode rdf:resource="#A_CN5" />
  </cim:Terminal>
  <cim:ConnectivityNode rdf:ID="A_CN5"></cim:ConnectivityNode>
  <cim:Terminal rdf:ID="A_T14">
    <cim:Terminal.ConductingEquipment rdf:resource="#A_M" />
    <cim:Terminal.ConnectivityNode rdf:resource="#A_CN5" />
  </cim:Terminal>
  <cim:BusbarSection rdf:ID="A_M">
    <cim:IdentifiedObject.name>Busbar M</cim:IdentifiedObject.name>
    <cim:IdentifiedObject.mRID>30</cim:IdentifiedObject.mRID>
  </cim:BusbarSection>
  <cim:Terminal rdf:ID="A_T15">
    <cim:Terminal.ConductingEquipment rdf:resource="#A_D" />
    <cim:Terminal.ConnectivityNode rdf:resource="#A_CN5" />
  </cim:Terminal>
  <cim:ACLineSegment rdf:ID="A_D">
    <cim:Conductor.length>2000</cim:Conductor.length>
    <cim:Conductor.r>0.25</cim:Conductor.r>
    <cim:Conductor.x>0.224</cim:Conductor.x>
    <cim:Conductor.bch>188.496</cim:Conductor.bch>
    <cim:IdentifiedObject.name>Line D</cim:IdentifiedObject.name>
    <cim:IdentifiedObject.mRID>32</cim:IdentifiedObject.mRID>
  </cim:ACLineSegment>
  <cim:Terminal rdf:ID="A_T4">
    <cim:Terminal.ConductingEquipment rdf:resource="#A_D" />
    <cim:Terminal.ConnectivityNode rdf:resource="#A_CN1" />
    <cim:IdentifiedObject.mRID>33</cim:IdentifiedObject.mRID>
  </cim:Terminal>
</rdf:RDF>
```

Appendix F
Description of Message Type Verbs

The following table shows commonly used Verbs in IEC 61968 (originally taken from IEC 61968-9 [57])

Proposed Verbs	Meaning	Message Body
CREATE	The CREATE verb is used to submit a request to the master system to create a new document. The master system may in turn publish the new document using the verb CREATED. The master system may also use the verb REPLY to response to the CREATE request, indicating whether the request has been processed successfully or not.	Header, Payload
UPDATE	The UPDATE verb is used to submit a request to the master system to make a change in the document based on the information in the message. The master system may in turn publish the changed document using the verb UPDATED to notify that the document has been changed since last published. The master system will use the verb REPLY to response to the UPDATE request, indicating whether the request has been processed successfully or not.	Header, Payload
CHANGE	Synonym for UPDATE	Header, Payload

CANCEL	The CANCEL verb is used to submit a request to the master system to cancel the document. The master system may in turn publish the cancelled message using the verb CANCELED to notify that the document has been cancelled since last published. The master system will use the verb REPLY to respond to the CANCEL request, indicating whether the request has been processed successfully or not. The CANCEL verb is used when the business content of the document is no longer valid due to error(s).	Header, Request
CLOSE	The CLOSE verb is used to submit a request to the master system to close the document. The master system may in turn publish the closed message using the verb CLOSED to notify that the document has been closed since last published. The master system will use the verb REPLY to response to the CLOSE request, indicating whether the request has been processed successfully or not. The CLOSE verb is used when the business document reaches the end of its life cycle due to successful completion of a business process.	Header, Request
DELETE	The DELETE verb is used to submit a request to the master system to delete the document. The master system may in turn publish the closed message using the verb DELETED to notify that the document has been deleted since last published. The master system may also use the verb REPLY to response to the DELETE request, indicating whether the request has been processed successfully or not. The DELETE verb is used when the business document should no longer be kept in the integrated systems either due to error(s) or due to archiving needs.	Header, Request

GET	The GET verb is used to submit a query request to the master system to get the current data for a given document reference code or a set of documents. The master system will use the verb REPLY to respond to the GET request.	Header, Request
CREATED	The CREATED verb is used to publish the creation of a document as a result of either an external request or an internal action within the master system of that document. This is the first time that data for this document reference code has been published as the result of internal or external request; in which case, it would use the same document reference as the CREATE message. This message type is usually subscribed by interested systems and could be used for mass updates. There is no need to reply to this message type.	Header, Payload
UPDATED	The UPDATED verb is used to publish events related to the change of a document as a result of either an external request or an internal action within the master system of that document. This could be a generic change in the content of the document or a specific status change such as "‘approved"’, "‘issued"’ etc. This message type is usually subscribed by interested systems and could be used for mass updates. There is no need to reply to this message type.	Header, Payload
CHANGED	Treat as a synonym with UPDATED	Header, Payload
CLOSED	The CLOSED verb is used to publish events related to the normal closure of a document as a result of either an external request or an internal action within the master system of that document. This message type is usually subscribed by interested systems and could be used for mass updates. There is no need to reply to this message type.	Header, Payload

CANCELED	The CANCELED verb is used to publish the cancellation of a document as a result of either an external request or an internal action within the master system of that document. This message type is usually subscribed by interested systems and could be used for mass updates. There is no need to reply to this message type.	Header, Payload
DELETED	The DELETED verb is used to publish the deletion of a document as a result of either an external request or an internal action within the master system of that document. This message type is usually subscribed by interested systems and could be used for mass updates. There is no need to reply to this message type.	Header, Payload
SHOW	Synonym to REPLY, may be deprecated.	Header, Reply, Payload
REPLY	The REPLY verb is used to return the processing result of a request for a CREATE, UPDATE, DELETE, CANCEL or CLOSE.	Header, Reply, Payload
SUBSCRIBE	The SUBSCRIBE verb is used to indicate a subscription for a type of information identified by the noun. This is realized within the integration infrastructure (e.g. JMS).	Not implemented
UNSUBSCRIBE	The UNSUBSCRIBE verb is used to indicate the termination of a subscription for a type of information identified by the noun. This is realized within the integration infrastructure (e.g. JMS).	Not implemented

Glossary

API. An application programming interface (API) is a particular interface provided to allow other software systems to access functionality of a specific system. A potential API defines, for instance, the data types that can be exchanged, dedicated functionality and how it can be used in context.

Advanced Metering Infrastructure (AMI). Advanced Metering Infrastructure are systems that typically measure, collect and analyze energy (most of the times, gas, heat or electricity) usage, and communicate with metrological devices such as electricity, gas, heat, and water meters, either on request, real-time or on a fixed schedule (which can differ from quarter of an hour to annual). The systems include hardware, software, communications, consumer energy displays and controllers, customer associated systems, Meter Data Management (MDM) software, and supplier business systems.

CENELEC. CENELEC (French: Comité Européen de Normalisation Électrotechnique) is the European Committee for Electrotechnical Standardization. It is responsible for European Standardization in the area of electrical engineering. Together with the ETSI (telecommunication) and CEN (other technical areas), CENELEC forms the European system for Standardization.

CIM users Group (CIMug). The CIM Users Group is dedicated to the promotion of the portability of existing applications and to the promotion of the ease of installation of new applications through the use of such standards as the common information model, message bus, and common data access specification. The CIM Users Group is under the administrative umbrella of the UCA International Users Group, a not-for-profit corporation.

Combined Heat and Power (CHP). Cogeneration (also called combined heat and power, CHP) is the use of a heat engine or a power station to simultaneously generate both electricity and useful heat. All power plants emit a certain amount of heat during electricity generation. This can be released into the natural environment through cooling towers, flue gas, or by other means. By contrast, CHP captures the by-product heat for heating purposes, either very close to the plant, or as hot water for district heating with temperatures ranging from approximately 80 to 130 degree Celsius. Small CHP plants are an example of decentralized energy resources.

COmpanion Specification for Energy Metering (COSEM). COSEM or Companion Specification for Energy Metering, includes a set of specifications that defines the Transport and Application Layers of the DLMS protocol.

Cyber Security. Cyber security is a branch of computer technology also known as Information Security as applied to computers and networks. The objective of cyber security includes protection of information and property from theft, corruption, or natural disaster, while allowing the information and property to remain accessible and productive to its intended users.

Decentralized Energy Resources (DER). Distributed generation, also called on-site generation, dispersed generation, embedded generation, decentralized generation, decentralized energy or distributed energy, generates electricity from many small energy sources.

Demand response. In electric smart grids, demand response (DR) is similar to dynamic demand mechanisms to manage customer consumption of electricity in response to supply conditions, for example, having electricity customers reduce their consumption at critical times or in response to market prices.

Device Language Message specification (DLMS). DLMS or Device Language Message Specification (originally Distribution Line Message Specification), is the suite of standards developed and maintained by the DLMS User Association and has been co-opted by the IEC TC13 WG14 into the IEC 62056 series of standards.

Distribution grid. The transport of generator-produced electric energy to loads is done by the grid. An electric power transmission system interconnects generators and loads and generally provides multiple paths among them. Multiple paths increase system reliability because the failure of one line does not cause a system failure. With operating voltages less than 34.5 kV, the distribution system carries energy from the local substation to individual households, using both overhead and underground lines.

Distribution system operator(DSO). A natural or legal person responsible for operating, ensuring the maintenance of and, if necessary, developing the local distribution system in a given area and, where applicable, its interconnections with other systems and for ensuring the long-term ability of the system to meet reasonable demands for the distribution of electricity. Moreover, the DSO is in most regulatory regimes responsible for regional grid access and grid stability, integration of renewables at the distribution level and regional load balancing.

DMS. The term distribution management system coins an energy management system which has to optimize, track and run the operations and optimizations in the so called distribution grid which has to be operated different than the so called transmission grid.

Domain ontology. A domain ontology (or domain-specific ontology) models a specific domain, which represents part of the world. Particular meanings of terms applied to that domain are provided by domain ontology.

E-Energy. E-Energy - Smart Grids made in Germany is a funding program of the Federal Ministry of Economics and Technology (BMWi) in an inter-ministerial partnership with the Federal Ministry for the Environment, Nature Conservation and Nuclear Safety (BMU). Climate change, the rapid surge in energy demand and dwindling natural resources present Germany with major challenges in the field of energy supply. Industry and politics must work hand in hand in future to secure an economical and environmentally compatible supply of power for all public and private sectors.

Electric Power Research Institute (EPRI). The Electric Power Research Institute (EPRI) conducts research on issues related to the electric power industry in USA. EPRI is a nonprofit organization funded by the electric utility industry. EPRI is primarily a US based organization, receives international participation. EPRI's area covers different aspects of electric power generation, delivery and its use.

Electric Vehicles (EV). An electric vehicle (EV), also referred to as an electric drive vehicle, uses one or more electric motors or traction motors for propulsion. Three main types of electric vehicles exist, those that are directly powered from an external power station, those that are powered by stored electricity originally from an external power source, and those that are powered by an on-board electrical generator, such as an engine (a hybrid electric vehicle), or a hydrogen fuel cell.

Enterprise application integration (EAI). Enterprise Application Integration (EAI) is a concept to integrate business related functions across enterprises using software and computer systems. This involves an architectural approach planning required applications and considering for instance data and business processes.

EMS. An energy management system (EMS) is a system of computer-aided tools used by operators of electric utility grids to monitor, control, and optimize the performance of the generation and/or transmission system. The monitor and control functions are known as SCADA; the optimization packages are often referred to as "advanced applications".

Enterprise Resource Planning (ERP). Enterprise resource planning (ERP) integrates internal and external management information across an entire organization, embracing finance/accounting, manufacturing, sales and service, customer relationship management, etc. ERP systems automate this activity with an integrated software application.

ESB. An enterprise service bus (ESB) is a software abstraction layer used to integrate usually distributed and heterogeneous services in Service Oriented Architectures. It is usually based on message-oriented solutions and provides for example routing mechanisms, data transformation and services registries.

Flexible Alternating Current Transmission System (FACTS). A flexible alternating current transmission system (FACTS) is a system composed of static equipment used for the AC transmission of electrical energy. It is meant to enhance controllability and increase power transfer capability of the network. It is generally a power electronics-based system.

Geographic information system (GIS). Geographic information system (GIS), geographical information system, or geospatial information system are systems designed to capture, store, manipulate, analyze, manage, and present all types of geographically referenced data. In the simplest term, GIS is the merging of cartography, statistical analysis, and database technology.

High-Voltage Direct Current (HVDC). A high-voltage, direct current (HVDC) electric power transmission system uses direct current for the bulk transmission of electrical power, in contrast with the more common alternating current systems. For long-distance transmission, HVDC systems may be less expensive and suffer lower electrical losses.

Home automation. Home automation is the residential extension of "industrial building automation". It is automation of the home, housework or household activity. Home automation may include centralized control of lighting, HVAC (heating, ventilation and air conditioning), appliances, and other systems, to provide improved convenience, comfort, energy efficiency and security.

IEC. The International Electrotechnical Commission (IEC) is a non-profit, non-governmental international standards organization that prepares and publishes International Standards for all electrical, electronic and related technologies ? collectively known as "electrotechnology".

IEEE. The Institute of Electrical and Electronics Engineers (IEEE) is a non-profit professional association dedicated to advancing technological innovation related to electricity.

Information and Communication Technologies (ICT). ICT is used as a general term for all kinds of technologies which enable users to create, access and manipulate information. ICT is a combination of information technology and communications technology. In an increasingly interconnected world, the interactions among devices, systems, and people are growing rapidly.

Intelligent Electronic Device (IED). An Intelligent Electronic Device (IED) is a term used in the electric power industry to describe microprocessor-based controllers of power system equipment, such as circuit breakers, transformers, and capacitor banks.

Internet of Energy. As with the Internet of Things, also the corresponding concept of the Internet of Energy has been coined by various stakeholders being the combination of ICT and field automation for power grid management.

ISO. The International Organization for Standardization, widely known as ISO, is an international standard-setting body composed of representatives from various national standards organizations. Founded on February 23, 1947, the organization promulgates worldwide proprietary industrial and commercial standards.

Legacy processes. Processes are considered legacy processes if new systems using EAI technologies must be integrated with those processes in terms of function coupling at migration level.

Load. If an electric circuit has a well-defined output terminal, the circuit connected to this terminal (or its input impedance) is the so called load.

Microgrid. A microgrid is a often considered a localized grouping of both electricity generation and energy storage, alongside loads that normally operates connected to a traditional, more centralized grid (so called macrogrid). This single point of common coupling with the macrogrid can be disconnected without interrupting supply security. The microgrid can function autonomously afterwards.

OPC UA. OPC Unified Architecture is the most recent OLE for process control (OPC) specification from the OPC Foundation and differs significantly from its predecessors.

Peak load. In the United States, this often occurs in the afternoon, especially during the summer months when the air conditioning load is high. The peak power load generally occurs when people return home from work, start cooking dinner, and turn up the air conditioning. During this time many workplaces are still open and consuming power.

Power transmission. Power transmission is the movement of energy from its place of generation to a location where it is applied to performing useful work.

Prosumer. Prosumer is a word formation consisting of the word professional or sometimes producer and the word consumer.

RDF. The Resource Description Framework (RDF) is defined by a set of W3C recommendations and is originally intended to serve as a standard model for meta data interchange on the world wide web using a graph-based structure. RDF data can be expressed in various formats, as e.g. XML, Turtle or N3.

Regulation. Regulation is administrative legislation that constitutes or constrains rights and allocates responsibilities. It can be distinguished from primary legislation (by Parliament or elected legislative body) on the one hand and judicial decisions on the other hand.

Requirements. In engineering, a requirement is a singular documented need of what a particular product or service should be or perform. It is most commonly used in a formal sense in systems engineering, software engineering, or enterprise engineering. It is a statement that identifies a necessary attribute, capability, characteristic, or quality of a system in order for it to have value and utility to a user.

SIA. The Seamless Integration Architecture SIA by the IEC TC 57 is the general architecture picture taking into account the layered model of existing and future IEC TC 57 standards.

SCADA. SCADA (supervisory control and data acquisition) generally refers to industrial control systems: computer systems that monitor and control industrial, infrastructure, or facility-based processes

SOA. Service-oriented architecture (SOA) is a flexible set of design principles used during the phases of systems development and integration in computing. A system based on a SOA will package functionality as a suite of interoperable services that can be used within multiple, separate systems from several business domains.

SOAP. Once known as Simple Object Access Protocol, SOAP is no longer an acronym. SOAP is used as an envelop within web services for payload delivery.

Smart Grid. A Smart grid is a type of electrical grid which attempts to predict and intelligently respond to the behaviour and actions of all electric power users connected to it - suppliers, consumers and those that do both ? in order to efficiently deliver reliable, economic, and sustainable electricity services.

Substation Automation. Substation automation refers to using data from Intelligent electronic devices (IED), control and automation capabilities within the substation, and control commands from remote users to control power system devices.

TC. A TC is a so called technical committee within IEC which is a group formed by IEC with corresponding subgroups which are reponsible for created certain standards.

Tele-control. The IEC 60870 part 5 provides a communication profile for sending basic tele-control messages between two systems, which uses permanent directly connected data circuits between the systems.

Transmission grid. The transport of generator-produced electricity to loads is done by a grid. An electric power transmission system connects generators and loads and, generally, provides multiple paths for routing among them. Multiple paths increase system reliability because the failure of one line does not cause a system failure.With operating voltage exceeding 230 kV, the transmission system interconnects generating stations and large substations located close to load centers by using overhead lines.

Transmission system operator(TSO). A transmission system operator (TSO) is an entity entrusted with transporting energy in the form of natural gas, heat and/or electrical power on a national or regional level, using a fixed, often un-bundled infrastructure.

UCA. The UCAIug as well as its member groups (CIMug, OpenAMI, and IEC61850) draws its membership from utility user and supplier companies. The mission of the CIMug is to manage and to communicate issues concerning the CIM model and to serve as the primary means for developing the CIM model consensus and consistency across the industry.

UML. The Unified Modeling Language (UML) is a standardized, formal, object-oriented modeling language, used to describe software-intensive systems. It includes the definition of graphical representations to create and represent the underlying model, and is specified and managed by the Object Management Group.

Vehicle-to-grid (V2G). The term Vehicle-to-grid (V2G) describes a system in which plug-in electric vehicles (EV), such as electric cars (BEVs) and plug-in hybrids (PHEVs), communicate with the grid to sell demand response services by either delivering electricity into the grid or throttling their charging rate.

XML. The Extensible Markup Language (XML) is defined by the Consortium (W3C) and defines a formal grammar and vocabulary for structured, textual descriptions that are machine-readable as well as readable by humans. It consists of several specifications that are publicly available from the W3C website.

References

1. JTC 1 Special Working Group on Smart Grid (SWG-Smart Grid) (2010), `http://www.jtc1smartgrid.org/`
2. Altova: XML Editor, Data Management, UML, and Web Services Tools from Altova (2011), `http://www.altova.com/`
3. Britton, J.: Designing Model Exchange Processes with CIM and 'RMA Sets'. In: 2006 IEEE PES Power Systems Conference and Exposition, pp. 487–489 (2006), `http://ieeexplore.ieee.org/lpdocs/epic03/wrapper.htm?arnumber=4075799`, doi:10.1109/PSCE.2006.296360
4. Burke, T.J.: OPC and Intro OPC UA (2008)
5. CEN, CENELEC, ETSI: JWG Report on Standards for the Smart Grid. Tech. rep. (2010)
6. CIMug: CIM users group Website, `http://cimug.ucaiug.org`
7. CIMug: Flyer: About the CIM users group (2008)
8. D2.24: EMS Architectures for the 21st Century. Tech. rep. (2009)
9. Demaree, K.: TC 57 CIM Model Manager Report (2011)
10. DeVos, A.: RDF Difference Models; Representing the Difference between two RDF Models (2002)
11. DIN: Die deutsche Normungsstrategie aktuell (2009), `http://www.din.de/sixcms_upload/media/2896/DNS_2010d_akt.pdf`
12. DKE: The German Standardization Roadmap E-Energy/Smart Grid. VDE (2010), `http://www.dke.de/de/std/KompetenzzentrumE-Energy/Seiten/Links.aspx`
13. DKE: The German Standardization Roadmap E-Energy/Smart Grid - a First Update and Report. VDE (2011), `http://www.dke.de/de/std/KompetenzzentrumE-Energy/Seiten/Links.aspx`
14. Effantin, C.: Available List of Tools and their position in TC57 modelling stack (2010)
15. EPRI: Harmonization of IEC 61970, 61968, and 61850 Models (2006)
16. EPRI: IntelliGrid Architecture Application Guide: Metering and Consumer Systems (2006)
17. EPRI: Integration of Advanced Automation and Enterprise Information Infrastructures: Harmonization of IEC 61850 and IEC 61970/61968 Models (2007)
18. EPRI: An Introduction to the CIM for Integrating Distribution (2008)

19. EPRI: Report to NIST on the Smart Grid Interoperability Standards Roadmap. Tech. rep., EPRI (2009)
20. EPRI: Development of the Common Information Model for Distribution and A Survey of Adoption (2010)
21. EPRI: Harmonizing the International Electrotechnical Commission Common Information Model (CIM) and 61850 (2010)
22. González, J.M.: Presentation SMB SG3 Use Cases and Profiles - Update (2010)
23. Hitzler, P., Krötzsch, M., Rudolph, S.: Foundations of Semantic Web Technologies. Chapman and Hall/CRC (2009), http://www.amazon.com/Foundations-Semantic-Technologies-Textbooks-Computing/dp/142009050X
24. IEC: Smart Grid: Global Standards for optimal electricity delivery, http://www.iec.ch/smartgrid/
25. IEC: Smart Grid Mapping tool, http://www.iec.ch/smartgrid/mappingtool/
26. IEC: 61970-450 Energy Management System Application Program Interface (EMS-API) - Part 450: CIS Information Exchange Model Specification Guide (2002)
27. IEC: 61968-3 Ed. 1: Application integration at electric utilities - System interfaces for distribution management - Part 3: Interface for network operations (2003)
28. IEC: 61970-2 Ed.1: Energy management system application program interface (EMS-API) - Part 2: Glossary (2003)
29. IEC: 61970-451 Energy Management System Application Program Interface (EMS-API) - Part 451: CIS Information Exchange Model For SCADA (2003)
30. IEC: 62325-101 DTR Ed.1: Framework for energy market communications - Part 101: General guidelines and requirements (2004)
31. IEC: 62325-102 DTR Ed.1: Framework for energy market communications - Part 102: Energy market example model (2004)
32. IEC: 62325-501 DTR Ed.1: Framework for energy market communications - Part 501: General guidelines of using ebXML (2004)
33. IEC: 62325-502 DTS Ed.1: Framework for energy market communications - Part 502: Profile of ebXML (2004)
34. IEC: 61970-1 Ed.1: Energy management system application program interface (EMS-API) - Part 1: Guidelines and general requirements (2005)
35. IEC: 61970-401 Ed.1: Energy management system application program interface (EMS-API) - Part 401: Component interface specification (CIS) framework (2005)
36. IEC: 62351-6 TS Ed.1:Data and Communication Security - Part 6: Security for IEC 61850 Profiles (2005)
37. IEC: 61970-454 Energy Management System Application Program Interface (EMS-API) - Part 454: Naming Service Specification (2006)
38. IEC: 61970-501 Ed.1.0 Energy management system application program interface (EMS-API) - Part 501: Common information model resource description framework (CIM RDF) Schema (2006), http://www.thieme-connect.de/DOI/DOI?10.1055/s-2007-985912, doi:10.1055/s-2007-985912
39. IEC: 62351-1 TS Ed.1: Data and communication security - Part 1: Introduction and overview (2006)
40. IEC: 62351-3 TS Ed.1: Data and communication security - Part 3: Profiles including TCP/IP (2006)
41. IEC: 61968-1: Application integration at electric utilities - System interfaces for distribution management - Part 1: Interface architecture and general requirements (2007)

42. IEC: 61968-4 Ed.1: Application integration at electric utilities - System interfaces for distribution management - Part 4: Interfaces for records and asset management (2007)
43. IEC: 61970-301 Ed. 1: Energy management system application program interface (EMS-API) - Part 301: Common information model (CIM) base (2007)
44. IEC: 61970-404 Ed.1: Energy management system application program interface (EMS-API) - Part 404: High speed data access (HSDA) (2007)
45. IEC: 61970-405 Ed.1: Energy management system application program interface (EMS-API) - Part 405: Generic eventing and subscription (GES) (2007)
46. IEC: 61970-407 Ed.1: Energy management system application program interface (EMS-API) - Part 407: Time series data access (TSDA) (2007)
47. IEC: 61970-453 Ed.1: Energy management system application program interface (EMS-API) - Part 453: CIM based graphics exchange (2007)
48. IEC: 62351-2 Ed.1: Data and Communication Security - Part 2: Glossary of terms (2007)
49. IEC: 61968-13 Ed.1: Application integration at electric utilities - System interfaces for distribution management - Part 13: CIM RDF Model exchange format for distribution (2008)
50. IEC: 61968-6 Application integration at electric utilities - System interfaces for distribution management - Part 6: Interface Standard for Maintenance and Construction (2008)
51. IEC: 61970-402 Ed.1: Energy management system application program interface (EMS-API) - Part 402: Common services (2008)
52. IEC: 61970-403 Ed.1: Energy management system application program interface (EMS-API) - Part 403: Generic data access (2008)
53. IEC: 61970-502-8 Ed.1: Energy Management System Application Program Interface (EMS-API) - Part 502-8: CIM Data Services (2008)
54. IEC: 62351-7 TS Ed. 1: Power systems management and associated information exchange - Data and communication security - Part 7: Network and system management (NSM) data object models (2008)
55. IEC: 62541-1 Ed. 1.0: OPC Unified Architecture Specification - Part 1: Overview and Concepts (2008)
56. IEC: 61968-11 Ed.1: Application integration at electric utilities - System interfaces for distribution management - Part 11: Common Information Model (CIM) Extensions for Distribution (2009)
57. IEC: 61968-9 Ed.1: Application integration at electric utilities - System interfaces for distribution management - Part 9: Interface for meter reading and control (2009), `http://www.thieme-connect.de/DOI/DOI?10.1055/s-2007-985912`, doi: 10.1055/s-2007-985912
58. IEC: 61970-452: Energy Management System Application Program Interface (EMS-API) - Part 452: CIM Transmission Network Model Exchange Profile (2009)
59. IEC: 61970-552-4: CIM XML Model Exchange Format (2009)
60. IEC: 62351-7 TS Ed.1: Power systems management and associated information exchange - Data and communication security - Part 7: Network and system management (NSM) data object models (2009)
61. IEC: 62357 Second Edition: TC 57 Architecture - Part 1: Reference Architecture for TC 57 - Draft (2009)
62. IEC: 61968-1 System Inferfaces For Distribution Management - Part 1: Interface Architecture and General Recommendations (2010)
63. IEC: 61968-14 System Interfaces For Distribution Management - XML Naming and Design Rules (2010)

64. IEC: 61970-456 Energy Management System Application Program Interface (EMS-API) - Part 456: Solved Power System State Profiles (2010)
65. IEC: 62325-450 Deregulated Energy Market Communication - Part 450: Profile and Context Modelling Rules (2010)
66. IEC: 62361-100 Ed. 1.0 Harmonization of Quality Codes across TC 57 - Part 100: Naming and design rules for CIM profiles to XML schema mapping (2010)
67. IEC: 61968-1-1 Implementation Profile for IEC 61968 (2011)
68. IEC: 61968-100 (Draft): Application integration at electric utilities - System interfaces for distribution management - Part 100: Implementation Profiles for IEC 61968 (2011)
69. IEC: 61968-14 (Draft): Application integration at electric utilities - System interfaces for distribution management - Part 14: MultiSpeak - CIM Harmonization (2011)
70. IEC: 61968-2 Application integration at electric utilities - System interfaces for distribution management - Part 2: Glossary (2011)
71. IEC: 61968-8 Application integration at electric utilities - System interfaces for distribution management - Part 8: Interface Standard for Customer Support (2011)
72. IEC: 62351-8 Ed. 1.0 Power systems management and associated information exchange - Data and communications security - Part 8: Role-based access control (Draft) (2011)
73. IEC: 62351-9 Power systems management and associated information exchange - Data and communications security - Part 9: Cyber security key management for power system equipment (NWIP) (2011)
74. IEC: IEC 62325: Framework for Energy Market Communications - Part 351: CIM European Market Model Exchange Profile (2011)
75. IEC: IEC International Standards (IS) (2011), `http://www.iec.ch/standardsdev/publications/`
76. IEC: International Electrotechnical Commission (IEC) (2011), `http://www.iec.ch`
77. IEEE: P2030 Draft Guide for Smart Grid Interoperability of Energy Technology and Information Technology Operation with the Electric Power System (EPS). Tech. rep. (2009)
78. International Standardization Organization (ISO): ISO 8879:1986(E). Information processing âĂŤ Text and Office Systems âĂŤ Standard Generalized Markup Language (SGML). Tech. Rep. 1, Geneva (1986)
79. ISO, IEC: ISO / IEC Directives Part 1: Procedures for the technical work (2011)
80. ISO, IEC: ISO / IEC Directives Part 2: Rules for the structure and drafting of International Standards (2011)
81. Japanese METI: International Standardization Roadmap for Smart Grid (2010)
82. Josuttis, N.M.: SOA in Practice, vol. 253. O'Reilly (2007)
83. JWG SG: Smart grid Mandate: Standardization Mandate to European Standardisation Organisations (ESOs) to support European Smart grid Deployment. Tech. rep., European Comission, Brussels (2011)
84. Kostic, T., Frei, C., Preiss, O., Kezunovic, M.: Scenarios for data exchange using standards IEC 61970 and IEC 61850. In: UCA User Group meeting, Cigre Paris, pp. 1–5 (2004), `http://scholar.google.com/scholar?hl=en&btnG=Search&q=intitle:Scenarios+for+Data+Exchange+using+Standards#0`
85. Kostic, T., Preiss, O., Frei, C.: Towards the Formal Integration of Two Upcoming Standards. IEC 61970 and IEC 61850. In: Little, E.H. (ed.) Proceedings fo the 2003 LESCOPE Conference, Montreal, May 7-9. IEEE Publishing (2003)
86. Lange, J., Iwanitz, F., Burke, T.J.: OPC: Von Data Access bis Unified Architecture. VDE (2010)

87. Mahnke, W., Leitner, S.H., Damm, M.: OPC Unified Architecture (2009), http://books.google.com/books?id=de9uLdXKj1IC\&pgis=1
88. Maizener, A., Sanson, J.l.: CimSyntaxGen Add-In User Guide V2 (2011)
89. Maligue-Clausse, S., Maizener, A., Sanson, J.L.: CimConteXtor User Guide V2 (2011)
90. National Institute for Standards and Technology: The NIST Definition of Cloud Computing (Draft). Tech. rep. (2011)
91. NIST: Guidelines for Smart Grid Cyber Security: Smart Grid Cyber Security Strategy, Architecture, and High-Level Requirements Vol 1,Tech. Rep. (August 2010)
92. NIST: NIST Framework and Roadmap for Smart Grid Interoperability Standards (2010)
93. NIST Smart Grid Collaboration: Priority Action Plans, http://collaborate.nist.gov/twiki-sggrid/bin/view/SmartGrid/PriorityActionPlans
94. Østergaard, J.: European SmartGrids Technology Platform-Vision and Strategy for Europes Electricity Networks of the Future (2006), http://scholar.google.com/scholar?hl=en&btnG=Search&q=intitle:European+SmartGrids+Technology+Platform#3
95. Object Management Group: OMG Unified Modeling Language (OMG UML), Infrastructure (2010), http://www.omg.org/spec/UML/2.3/Infrastructure/PDF/
96. Object Management Group: OMG Unified Modeling Language (OMG UML), Superstructure (2010), http://www.omg.org/spec/UML/2.3/Superstructure/PDF/
97. OFFIS and SCC Consulting and MPC management coaching: Untersuchung des Normungsumfeldes zum BMWi-Förderschwerpunkt 'E-Energy - IKT-basiertes Energiesystem der Zukunft' (2009)
98. Österlund, L.o.: WG13 CIM Model Manager Report (2010)
99. Pilone, D., Pitman, N.: UML 2.0 in a Nutshell (In a Nutshell (O'Reilly)). O'Reilly Media (2005), http://www.amazon.com/UML-2-0-Nutshell-OReilly/dp/0596007957
100. Postina, M., Trefke, J., Steffens, U.: An EA-approach to Develop SOA Viewpoints. In: 2010 14th IEEE International Enterprise Distributed Object Computing Conference, pp. 37–46. IEEE (2010), http://ieeexplore.ieee.org/xpl/freeabs_all.jsp?arnumber=5630229
101. Power Info: CIMSPY - Power Info - Utility IT Solution Provider (2011), http://www.powerinfo.us/index.html
102. Rohjans, S., Piech, K., Mahnke, W.: Standardized Smart Grid Semantics using OPC UA for Communication. IBIS - Interoperability in Business Information Systems 6(10) (2011)
103. Rohjans, S., Piech, K., Uslar, M., Cabadi, J.F.: CIMbaT - Automated Generation of CIM-based OPC UA-Address Spaces. In: IEEE SmartGridComm. 2011, Brussels (2011)
104. Rohjans, S., Uslar, M., Appelrath, H.J.: OPC UA and CIM: Semantics for the smart grid. In: Transmission and Distribution Conference and Exposition, 2010 IEEE PES, pp. 1–8 (2010)
105. Rohjans, S., Uslar, M., Bleiker, R., González, J., Specht, M., Suding, T., Weidelt, T.: Survey of Smart Grid Standardization Studies and Recommendations. In: First IEEE International Conference on Smart Grid Communications (2010)

106. Santodomingo, R., Rodriguez-Mondejar, J.A., Sanz-Bobi, M.A.: Ontology Matching Approach to the Harmonization of CIM and IEC 61850 Standards. IEEE (2010), http://ieeexplore.ieee.org/lpdocs/epic03/wrapper.htm?arnumber=5622017
107. SGCC: Framework and Roadmap for Strong and Smart Grid Standards (2009)
108. SMB Smart Grid Strategic Group (SG3): IEC Smart Grid Standardization Roadmap (2010)
109. Sommerville, I.: Software Engineering, 9th revise edn. Addison-Wesley Longman, Amsterdam (2010), http://www.amazon.de/Software-Engineering-Ian-Sommerville/dp/0137053460/ref=sr_1_1?ie=UTF8\&qid=1308056747\&sr=8-1
110. Specht, M., Osterloh, A.: Erweiterung des Common Information Model zur Modellierung von dezentralen Energieprodukten an einem regionalen Marktplatz. Informatik (2010)
111. The Open Group: TOGAF Version 9 - The Open Group Architecture Framework (TOGAF), 9th edn. (2009)
112. UN/CEFACT: XML Naming and Design Rules (2006), http://scholar.google.com/scholar?hl=en&btnG=Search&q=intit%le:XML+Naming+and+Design+Rules#1
113. Uslar, M.: Ontologiebasierte Integration heterogener Standards in der Energiewirtschaft. Ph.D. thesis, Edewecht (2010)
114. Uslar, M., Dahlem, N.: Semantic web technologies for power grid management. In: Koschke, R., Herzog, O., Rödiger, K.H., Ronthaler, M. (eds.) GI Jahrestagung (1). LNI, vol. 109, pp. 242–250. GI (2007)
115. Uslar, M., Rohjans, S., Bleiker, R., González, J.M., Suding, T., Specht, M., Weidelt, T.: Survey of Smart Grid Standardization Studies and Recommendations - Part 2. In: IEEE Innovative Smart Grid Technologies Europe (2010)
116. Uslar, M., Rohjans, S., Specht, M., Gonzales, J.: What is the CIM lacking? In: IEEE SmartGridComm 2010 (2010)
117. World Wide Web Consortium (W3C): RDF Primer (2004), http://www.w3.org/TR/rdf-primer/
118. World Wide Web Consortium (W3C): RDF Semantics (2004), http://www.w3.org/TR/rdf-mt/
119. World Wide Web Consortium (W3C): RDF Vocabulary Description Language 1.0: RDF Schema (2004), http://www.w3.org/TR/rdf-schema/
120. World Wide Web Consortium (W3C): Resource Description Framework (RDF): Concepts and Abstract Syntax (2004), http://www.w3.org/TR/rdf-concepts/
121. World Wide Web Consortium (W3C): XML Schema Part 0: Primer Second Edition (2004), http://www.w3.org/TR/xmlschema-0/
122. World Wide Web Consortium (W3C): XML Schema Part 1: Structures Second Edition (2004), http://www.w3.org/TR/xmlschema-1/
123. World Wide Web Consortium (W3C): XML Schema Part 2: Datatypes Second Edition (2004), http://www.w3.org/TR/xmlschema-2/
124. World Wide Web Consortium (W3C): Extensible Markup Language (XML) 1.1, 2nd edn. (2006), http://www.w3.org/TR/xml11/

125. World Wide Web Consortium (W3C): Namespaces in XML 1.1, 2nd edn. (2006), http://www.w3.org/TR/xml-names11/
126. World Wide Web Consortium (W3C): Extensible Markup Language (XML) 1.0, 5th edn. (2008), http://www.w3.org/TR/REC-xml/
127. World Wide Web Consortium (W3C): OWL 2 Web Ontology Language Primer (2009), http://www.w3.org/TR/owl-primer/
128. World Wide Web Consortium (W3C): OWL 2 Web Ontology Language RDF-Based Semantics (2009), http://www.w3.org/TR/owl-rdf-based-semantics/
129. World Wide Web Consortium (W3C): W3C RDF Validation Service (2011), http://www.w3.org/RDF/Validator/
130. Xtensible Solutions, Inc.: CIMEA Webpage (2011), www.cimea.org

Index

MIX
Papier aus verantwortungsvollen Quellen
Paper from responsible sources
FSC® C105338

If you have any concerns about our products,
you can contact us on
ProductSafety@springernature.com

In case Publisher is established outside the EU,
the EU authorized representative is:
Springer Nature Customer Service Center GmbH
Europaplatz 3, 69115 Heidelberg, Germany

Printed by Libri Plureos GmbH
in Hamburg, Germany